Goldfields of Northwestern Ontario

A History of the Patricia Transportation Company

By Jack Wish

Goldfields of Northwestern Ontario

Book packaged by Great Plains Publications Ltd., Winnipeg, Manitoba
Design by Taylor George Design, Winnipeg, Manitoba
Printed in Canada by Friesens, Altona, Manitoba

Canadian Cataloguing in Publication Data
Wish, Jack

Goldfields of Northwestern Ontario

Includes bibliographical references and index.
ISBN 0-9681207-0-9

1. Patricia Transportation Co.—History.
2. Transportation—Ontario, Northern—History.
3. Rural development—Ontario, Northern—History.
I. James Richardson & Sons. II. Title

HE199.C3W57 1996 388'.065713'11 C96-920114-I

Table of Contents

DEDICATION

This book is dedicated to the personnel of the Patricia Transportation Company, both those remaining with us and those who have passed on, who worked with the company during its 34 years of operation (1931-1964). Their work is a legacy to past and present generations of the Patricia region of Northwestern Ontario.

FOREWARD

BY HONOURABLE LEO BERNIER LLD

Here is a book that provides an excellent history of the pioneer spirit that enabled Canada to unlock the resources of its vast and rich Pre-Cambrian Shield. Jack Wish, in this history of the Patricia Transportation Company, vividly records the tremendous contribution made by all those involved in opening up a remarkable frontier area — Northwestern Ontario.

Many of those colourful characters of that area have since passed away or moved on; their children have grown and matured. But memories of that time will live on thanks to this author's initiative and dedication.

Through Jack Wish's efforts, he has captured history that I am sure may otherwise be lost. This will be an important resource for our children and future generations.

We, in this vast area of Northwestern Ontario, are truly grateful to Jack Wish for bringing to life and recording an exciting part of our past. As the pages of this book become worn from use and age, I will always consider it an honour to have contributed to this publication in a very small way.

PREFACE

THE LAST GREAT GOLD RUSH

By Christmas of 1925, the news that two Ontario prospectors had discovered a major gold find near Red Lake began leaking out into the national press. A month later, men from every walk of life, from all across Canada, were leaving their respective jobs and piling into CNR coach cars bound for the tiny railway town of Hudson, Ontario. Their slogan was "Red Lake or Bust".

Most of them did not have any idea where Hudson was located or what to expect upon arrival. Reality hit them when they got off the train in blowing snow and bone-chilling temperatures. They stood there in confusion, grasping the sobering reality that they now faced 135 miles of snow-covered lakes and portages before they even arrived in Red Lake to begin their search for a pot of gold.

Some came equipped with fur-lined parkas, tents, snowshoes, sleighs, toboggans and dog teams. Others came clad in suits and city shoes. As trains arrived with more men, accommodations were at a premium. Joe Kenneally's and Johnny Hiley's hotels were full to the rafters. Every spare room in the few private homes that existed was rented out. Joe Kenneally had a place in the back of his hotel where a man could sleep on a bunk or the floor for one dollar a night. The Hudson's Bay Company store had an Indian house at the back of their store where one could sleep for 25 cents a night. Others bedded down for free on the hard floor of the railway station's waiting room.

Those who had tents with necessary gear camped on the shore of Lost Lake where flames from camp fires flickered in the cold night. Meanwhile, in the town, queues formed outside restaurants and stores. Merchants wired for express shipments to take care of the extraordinary demand for warm clothing, food, sleighs, toboggans, snowshoes, axes and shovels, as well as sled dogs.

The trek into the frozen wilderness entailed at least six days and nights in subzero temperatures. At first Huskies were used, but by break-up every breed and description of dog in the area was in

Opposite page: Miners at the Beaver Mine in Northwestern Ontario c.1890.

Opposite page: Chute pulling at the 2050 foot level of the Central Patricia Gold Mines in 1941.

harness. Animals yelping, equipment clanking, the largely amateur prospectors soon set out northwards. Indian trappers marked out safe dog trails, which were packed as hard as a city street with the continuous marching of this new wave of explorers. Those who couldn't afford a dog team trudged the trail on snowshoes or dragged a toboggan loaded with their possessions and food supplies.

Bill Smith, Howard Halverson and Frank Aldens were the Hudson's Bay Company post managers at Red Lake, Goldpines and Lac Seul respectively. They had adequate food supplies for a normal winter's requirements, but soon realized that it was inadequate for all the men heading for Red Lake. In order to meet the demand, they hired teams of horses to bring in additional food supplies by sleighs to their posts. It was hard on the animals, as they had to work for many hours in bitter cold without protection from the weather. As a result, many perished from exposure.

It usually took one day to reach the HBC post on Lac Seul. There, for 25 cents a night, the men could sleep in the Indian House used by trappers. There they unpacked their food, cooked bannock and ate beans with bacon. Then they boiled pots of snow for tea and melted lard with cornmeal mush for their dogs.

The second day's travel brought them to Sand's Fish Camp near Windigo Point, which Dan Stewart had rented for the winter. He had three buildings where fishermen lived in the summer and Stewart turned this migration into a money-making enterprise. He served meals, brought in supplies of dog food and charged one dollar per night for all who wanted to bunk there. Late arrivals slept on the floor for the same price.

Travellers spent the third night at Pineridge (Goldpines). When the Indian House was full, men pitched their tents along the shore of Lac Seul. On the fourth and fifth nights of their journey, they had to pitch tents in the open bush country, before finally reaching Red Lake.

This tide of humanity flowed from Hudson to Red Lake until break-up in April. When the navigation season opened on Lac Seul, the prospectors continued to surge northward, but by boat instead of sleigh. By canoes or skiffs with outboard motors, they travelled the Chukuni River waterway to Red Lake, after crossing four portages en route.

I didn't personally arrive in Hudson until the gold rush was well under way, and the Patricia Transportation Company had already been formed. But I consider myself fortunate to have been part of the excitement. For the next 27 years, the Hudson area was to become the focal point for the last great gold rush in Canada. As an employee of the Patricia Transportation Company, I played a small part in supplying the prospectors and mine developers of this boom period in Northwestern Ontario. This is the story of that remote place and adventurous time.

Introduction

The World's Most Ancient Rock

Before this account of the North and its hidden riches can begin, we must first tell the story of a remarkable body of rock called the Canadian Shield.

The Canadian Shield is 2.5 billion years old, older than the mountains, the oceans, the plains or any of the earth's living creatures. It is the largest expanse of Precambrian rock in the world. "Precambrian" refers to that period of time predating the origins of life. Precambrian rock underlies all the rock on the surface of the earth and is, in most places, overlain with hundreds of feet of sedimentary rock and soil. Successive Ice Ages, each one lasting about 100,000 years, have removed the upper layers of material, exposing this ancient, mineral-rich Canadian Shield.

So much of Canada has been glaciated that almost half of the country's land mass is Precambrian Shield — about 4.7 million square miles, or an area of land larger than the entire subcontinent of India. This glaciated region covers 95 per cent of Quebec, 70 per cent of Ontario, 50 per cent of Manitoba, and about half of the Northwest Territories. Most of the Shield country is a wilderness populated only by animals. In fact, less than 10 per cent of Canada's human population inhabits the Shield. It is no mystery that only the hardiest people choose to live in the Shield. It is a land of stormy winters, voracious insects, vast swamps, deep forests and barren tundra where winter temperatures may plunge to 50° F below zero.

Historically, even aboriginal peoples preferred to settle in the milder zones to the south, where corn and other crops were easy to grow and wild animals were more plentiful. It wasn't until the glaciers retreated 10,000 years ago, however, that the Shield became home to these small, roving bands of Cree, Ojibway, Chipewyan and other First Nations who developed ingenious methods of wresting a living from this wilderness of rock, trees and water.

When European explorer Jacques Cartier

Opposite page: Frontier artist Paul Kane depicted this Shield scene on the Winnipeg River in 1846.

arrived in 1534, he took one look at Canada's vast, rocky Shield and pronounced it unfit for human habitation. "I did not find a cartload of earth though I landed in many places. In short...I deem that this is the land that God gave to Cain." Cartier did not foresee that one day the rivers of the Shield would become conduits of wealth for entrepreneurs and fur traders. Because the Shield is nearly devoid of soil, it is poorly drained and is laced with thousands of lakes and streams. After the last Ice Age, these waterways were inhabited by millions of beaver — an animal whose thick, glossy fur became the most fashionable material in Europe during the 17th and 18th centuries.

On May 2, 1670, King Charles II of England awarded Hudson Bay and its gigantic drainage basin to his cousin Prince Rupert and associates, naming them as the "true and absolute lords and proprietors" of a vast unexplored region known as Rupert's Land. By this act, the Hudson's Bay Company was created.. The first governor of the company was The Duke of York, and in his name York Factory was established at the mouth of the Hayes River to distribute supplies to other HBC posts. From these posts, fur traders encouraged local Indians to go out and trap the estimated 10 million beaver inhabiting Rupert's Land. From the early 1650's to the late 1850's, beaver became the breathing equivalent of gold. Men risked their lives and reputations, caught up in a feverish pursuit of the animal's lustrous fur. Beaver felt hats provided an elegant way to keep dry but, more importantly, made a fashion statement that the wearer was a person of wealth and influence.

To carry furs back and forth across the Shield, the traders made use of the system of rivers and lakes that interlaced the rocky wilderness. By using large canoes made from the bark of indigenous birch trees, they turned nature to their advantage and opened up a country that was thought by their forbears to be impenetrable. However, after 200 years of great drama and giant fortunes, the fur trade began to decline and the Shield country once again became a depopulated and desolate land. And so it would remain for almost a century until new types of wealth were discovered. It is the search for that new wealth, especially gold, that is the basis for the story I am about to tell.

Goldfields of Northwestern Ontario

A History of the Patricia Transportation Company

NORTHWESTERN
ONTARIO
SCALE 1:2,980,000 OR 47 MILES TO THE INCH
0
25
50
100
150
STATUTE MILES
0
25
50
100
150
KILOMETERS
HUDSON
Barton L.
Throat
Pikangikum Lake
Pikangikum
Berens
Cairns L.
Goose L.
Mamakwash Lake
Upper Goose L.
Kirkness L.
Nungesser Lake
Casummit Lake
Birch Lake
1276
Little Vermilion Lake
Swain Post
1294
Trout Lake
Narrow Lake
Cochenour
McKenzie Island
Red Lake
Madsen
Red
Snake Falls
Pakwash Lake
Bruce L.
Bluffy L.
Papaonga
Ear Falls
Goldpines
English
Anishinabi L.
Oak Lake
Maynard L.
105
Wabaskang L.
Perrault Falls
Perrault L.
Ball L.
English
Wabigoon
McKenzie Bay
Wapesi
Lac Seul
1170
Lac Seul
1156
Lount L.
McIntosh
Canyon
Clay L.
Cedar L.
Vermilion L.
Redditt
Jones
Niddrie
Amesdale
Quibell
Richan
Red Lake Road
Sunstrum
Silver L.
Edison
Kenora
Vermilion Bay
72
Hawk Lake
Pine
Eagle River
Oxdrift
Dryden
Wabigoon Lake
Dogtooth L.
Eagle Lake
1192
Wabigoon
71
Dryberry L.
Whitefish Bay
Sioux Narrows
Atikwa L.
Rowan Lake
Kakagi L.
Crow Lake
Nestor Falls
Pipestone L.
Loonhaunt L.
Burditt L.
Jackfish L.
Black Hawk
Rainy
Stratton
1108
Barwick
Emo
Mine Centre
Indus
Devlin
Fort Frances
Ranier
International Falls
Loman
Rainy
Black
VOYAGEURS
Cat Lake
Otoskwin
Kapikik L.
Gitche
Dobie L.
Dobie
Zionz L.
Fawcett L.
Gull L.
Pickle Lake
Pickle Crow
Central Patricia
Trading
Miminiska Lake
Albany
Fort Hope
Petawanga Lake
Attwood
Attwood Lake
Root Portage
Lake St. Joseph
Pashkokogan Lake
Greenbush L.
Grayson L.
Whiteclay L.
Whitewater Lake
Vincent L.
Hooker Lake
DeLesseps L.
Wabakimi L.
Ogoki Reservoir
1065
Control Dam
Savant Lake
1306
Smoothrock Lake
Caribou L.
599
Armstrong
Haystack Mt.
1400
1217
Sioux Lookout
Ghost River
Fowler
Savant Lake
Jacobs
Collins
Willet
Ferland
Ycliff
Allan Water
Ogaki
Green
Lamaune
Minataree
McDougall Mills
Kawaweogama Lake
Ombabika
Superior Junction
Zarn
Seseganaga Lake
Minnitaki Lake
Sturgeon Lake
Wapikaimaski Lake
Yonde
Umfreville
Harmon Lake
Obonga Lake
Lake
Watcomb (Wako)
Valora
Bell L.
Kashishibog L.
Press L.
Kukukus L.
Shikag L.
Metionga L.
855
Nipigon
Tannin
Paguchi L.
Holinshead L.
Gull
Sandybeach Lake
Basket L.
Dinorwic
Dyment
Barrel L.
Dinorwic L.
Raleigh
Gold Rock
Upper Manitou L.
Stormy Lake
Mameigwess L.
Ignace
17
Raleigh L.
Sowden Lake
Garden L.
Quorn
Wawang L.
Pakashkan L.
Poshkokagan
Lower Manitou L.
Bonheur
TRANS-CANADA HIGHWAY
English River
Graham
1619
Turtle
White Otter L.
Macdiarmid
Parks L.
Black Sturgeon L.
Pine Portage
1745
Beardmore
Clearwater West L.
Sandford L.
Upper Scotch L.
Lac des Îles
Frazer L.
McKirdy
11
Otukamamoan L.
Little Turtle
Seine
Upsala
Larson
Dog
Cameron Falls
Marmion Lake
Muskeg L.
Eaglehead L.
Spruce
Nipigon
Helen L.
Glenorchy
Crilly
1363
Steep Rock Lake
Savanne
James
Nipigon
Kama
TRANS-CANADA
1880
Seine
Calm L.
17
Red Rock
Lake
Flanders
Banning
Atikokan
Kawene
Quetico
1496
Lac des Mille Lacs
Raith
Dog Lake
1378
Hurkett
Dorion
Nipigon Bay
Pipestone
11
Huronian
Ouimet
1510
Quetico L.
Kashabowie
Griff
Pickerel
Namakan
Rossmere
Pearl
1320

Chapter One

Gold Fever in the North

The "Patricia" portion of the District of Kenora in Northwestern Ontario was named after the daughter of the Duke of Connaught, who was Canada's Governor General in 1912. This district is bounded in the north by Hudson Bay and by the province of Manitoba in the west. On the south, it follows the United States border through the western part of the Lake of the Woods to the Rainy River District. In the east, it is bounded by James Bay as far south as the Albany River, then west along the Albany (to the junction at 90° 48' longitude and 51° 0' latitude).

The Patricia District is a perfect example of the rough wilderness that Jacques Cartier described as "the land that God gave Cain". It has many lakes and rivers which run this way and that and make overland transportation exceedingly difficult. In 1804, Duncan Cameron wrote in *Les Bourgeois de la Compagnie du Nord-Ouest,* "The two thirds of this country are nothing but rivers and lakes, some fifty leagues long; properly speaking the whole country is nothing but water and islands; I have never travelled as yet above three leagues [about three miles] by land without finding a river or lake in my way."

The largest lakes are Lac Seul, Lake St. Joseph, Trout Lake and Lake of the Woods, while the major rivers are Albany, Attawapiskat, English, Wenasaga, Severn and Winisk. These lakes and rivers are teeming with fish such as northern pike, pickerel (walleye), muskie, suckers, tullibees and whitefish. As well, sturgeon are found in the Albany and Severn Rivers.

The land is covered with pine, spruce, balsam, tamarack, fir, birch, poplar, Norway pine and ash. Inhabiting the forest are bear, moose, deer, wolves, beaver, otter, fox, bobcat, ruffed grouse and many other smaller animals and birds. In the summer the bush is infested with mosquitoes, blackflies and a terrible little pest called the "no-see-um", so-called by the locals because they cannot be seen with a naked eye.

The first significant gold discovery in Northwestern Ontario was made at Jackfish Lake, in the Lake of the Woods region, in 1870. Soon there were a dozen gold producing mines there. In 1882, a Geological Survey of Canada report indicated that Red Lake had valuable mineral deposits. No actual finds were made until 1897 when a North Western Ontario Development Company team led by R.J. Gilbert hit paydirt. After paddling and portaging 200 miles, the party was thrilled to find gold lying on the rock surface near Slate Bay. They flaked off pieces and staked eight claims in the area before breaking camp. As they were about to depart, tragedy struck. Gilbert was picking up his revolver near the shore of the lake when it slid from the holster, hit the rock and discharged upward. Struck in the chest, Gilbert died instantly. Ironically, the early find at Slate Bay proved unproductive and the area was largely ignored for the next quarter century.

All the while, the provinces of Ontario and Manitoba were feuding over control of the mineral-rich region west of the Lakehead. Not until 1912 did Ontario finally prevail, when it was granted all the land up to the present day provincial boundary. Part of the new territory was named the Patricia District. There was hope in the older, more heavily populated area of southern Ontario that free land grants in the northwest would attract settlers. Despite the construction of the railway, however, the remoteness of the region and scarcity of arable land dissuaded most from relocating.

Even prospectors were not initially attracted to the region. Fantastic finds in eastern Ontario and the advent of the First World War meant there were few men willing or able to venture into the northwestern reaches of the province. Things changed in 1922 when a fur trader named George Swain took an ore sample he had found in an abandoned cabin to Winnipeg for analysis. In the series of events that followed, newspapers proclaimed the find to be a rich silver discovery. About 50 men made the trek to Red Lake and a mini rush was on. Silver prospectors soon found ribbons of more valuable gold, which then led to the major rush of 1926.

The Lure of Gold

Why search for gold? What is the lure of this rare metal? For more than 6,000 years, men have searched for gold, fought for it and founded civilizations on it. For most governments, gold is still the only universally accepted medium of exchange, the ultimate currency by which one nation settles its debts with another.

For centuries, gold has provided a benchmark for other forms of currency. Indeed, most paper currencies are based on the stability of a nation's gold supply. The United States' holdings at Fort Knox is the most famous example.

Gold's value is enhanced by its limited supply. There is so little gold in this world that all the 100,000 metric tons mined throughout history could be stacked in a cube measuring 60 feet each way — about the size of three family houses.

At the beginning of this century, mineral prospectors searched the Canadian Shield for gold, particularly concentrating on the northwestern part of the province of Ontario, an area of approximately 250,000 square miles. These gold prospectors experienced many hardships. They had to work in remote areas in very rough terrain and, most of the time, in inclement weather. They usually camped in small tents along the shores of lakes and rivers throughout the summer months from which they would go out early in the morning in hopes of finding gold. They returned to the camps as darkness fell, dead tired, with the hope that the next day would be more rewarding.

Prospecting also required funding. The initial outfitting was expensive, as were the recording fees required by government. Furthermore, claims had to be worked at least thirty days each year to be legally binding. Often, early excitement faded into

The Ancient Attraction

Gold's prime asset is its indestructibility. It does not tarnish (as silver does) and it resists corrosion. Gold coins have been recovered from sunken ships, after centuries beneath the sea, looking as bright as new. Primitive man found it almost as soft as putty, malleable and easy to work with. Cold, it could be hammered into magnificent masks like that of the Boy King who ruled Egypt 3,000 years ago. Over the centuries, skilled craftsmen learned they could beat a single ounce of gold into a sheet covering nearly 100 square feet. This could then be used as valuable decoration in everything from cathedrals to legislatures.

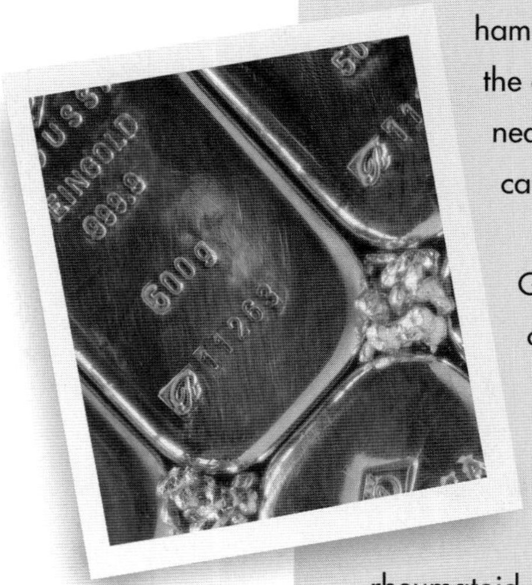

Great sculptors have also worked in gold. For example, Benvenuto Cellini made a solid gold salt cellar for Francis I of France four centuries ago. This salt shaker is in a museum today in Vienna and is still untarnished by age.

In medicine and folklore, there was widespread faith in gold's healing power. The most proven treatment was soluble gold salts injected in the muscles which eased the symptoms of rheumatoid arthritis. In India, powdered gold mixed with spices was used in many medicines.

Today, glass coated with a thin film of gold shrouds skyscrapers, not only to reflect the summer sun, but also to retain heat in winter. The use of such glass on one building in Edmonton, Alberta cut air-conditioning bills by 40 per cent. The ceiling of the Metropolitan Opera House in New York, the statue of Prometheus at Rockefeller Centre, and the domes of Colorado and Iowa's state capitals are all sheeted with gold.

In more practical terms, the modern electronics industry finds that nothing is as good as gold when it comes to conducting electricity, resisting corrosion and reflecting heat. It is indispensable in everything from pocket calculators to computers, telephones to television and missiles to spacecraft. The ordinary touch-tone telephone contains 33 gold contact points. Printed circuits of gold even guide the sequence of washing machines and dishwashers. "The electronics, telecommunications and computer industries as we know them would not exist without gold coatings," says the director of the Gold Institute in Washington.

disappointment as money ran out and the gold itself remained elusive. It wasn't long before corner posts and blazed bush trails were all that remained of a sudden burst of gold fever.

Sometimes, however, the prospectors got lucky and hit "paydirt". As I mentioned earlier, this happened in the summer of 1922 when a party of prospectors from Winnipeg discovered a vein of quartz containing silver-bearing galena near Red Lake. Press reports of this find attracted a number of prospectors, including Gus McManus who discovered small gold-bearing quartz stringers near the outlet of Red Lake.

Herbert Tyrell and Fred Carroll, prospecting at Red Lake, were sure they had found gold rather than silver. Ole Gustafson, Ole Sands and Wilfred Wright took Carroll up to Ear Falls in their fish boat *Triangle,* journeyed the rest of the way by canoe and feverishly staked claims in the area adjoining Carroll's. Following the 1922 rush of prospectors, each summer there was always a camp on

Prospector's tent north of Hudson in the winter of 1925.

the shore of Red Lake. The season of 1925 was no exception. Claims staked but not worked since 1922 had fallen open, and prospectors and geologists believed that gold must be hidden somewhere in the mass of Huronian rock.

On July 25, 1925, Lorne Howey and his partner George McNeeley discovered quartz stringers containing large amounts of native gold and staked a group of claims. Three posts would mark the location of what would later become the

Jack Hammell — Boxer, Sportswriter and Mine Promoter

John Edward Hammell was born on a farm in Beeton, Ontario in 1876. Interested in boxing, he pursued the sport in California under the name "Whirling Jack". After his career was cut short by an injury, Hammell moved into boxing promotion and sports writing in New York City. He met fellow Canadian boxer Jack Munro, and in 1905 the two joined the silver rush to Cobalt, Ontario. There Hammell worked variously as a guard and a bartender before moving into the risky business of buying and selling mining claims.

Hammell soon expanded into other mining ventures, including the 1907 silver rush to Gowganda, the 1909 Porcupine gold rush and the 1912 Kirkland gold rush. Hammell was also instrumental in securing financing for the huge Flin Flon, Manitoba copper-gold-zinc mine project in the 1920's.

Ever busy, Hammell was drawn to the Howey gold discovery of 1925, which triggered the famous Red Lake gold rush. He formed the Howey Red Lake Syndicate and optioned the claims to Dome Mines of South Porcupine. When Dome Mines failed to finish developing the property in 1926, Hammell formed Howey Gold Mines Limited. As president of its board of directors, he tried to bring the mine to fruition. However, the stock market crash of October 1929 nearly scuttled the project. But, board member W.S. Cherry, a wealthy Canadian merchant, stepped forward with a $500,000 loan offer. The mine began operating in 1930 and produced dividends by 1934.

Not one to relax, Hammell became president of Northern Aerial Mineral Explorations Company (NAME) in 1928, formed in co-operation with bush pilot and entrepreneur H.A. (Doc) Oaks. Hammell first met Oaks when he chartered planes from the Ontario Provincial Air Service to move men and supplies into Red Lake. Hammell and Oaks sent dozens of prospectors to work and, in 1928, it paid off when the Pickle Crow Gold Mine claim was staked. Hammell financed it into production and earned approximately $9 million in dividends over its 35 years of operation. NAME pilots covered more than one million miles in the exploration and development of mining in the previously uncharted Canadian north.

Hammell was president of several other lucrative gold mine operations, including Hasaga Gold Mines and Starratt Olson Gold Mines, both of Red Lake, as well as Uchi Gold Mines at Uchi Lake. He also lobbied vigorously for improvements to the town of Red Lake, including a highway connection to the outside world. Hammell died at the age of 82 in 1958, leaving behind an estate worth $2.5 million. So self-assured was he that an acquaintance of Hammell's recalled he used to sing his own variation of a well-known song: "The whole world is waiting for the sunrise...and Jack Hammell."

Company Characters

Ole Gustafson

Don Robinson, in the book *Golden Memories*, recalled an incident involving his father Carl and Ole Gustafson. Carl was celebrating Christmas Day in the company bunkhouse with Vic Nymark when the boss, Ole Gustafson, dropped over from his house about a hundred yards away.

Ole never drank and his wife had a reputation for frowning on those who did. Carl and Vic decided to get him drunk for a lark. They cajoled him into taking a Christmas drink and then another and another, each one larger than the last. They were enjoying it immensely, particularly when Ole's wife called them in for dinner. Ole walked off very well they thought, for a man who wasn't used to drinking. But they soon discovered why. When Vic pulled on his boots he found all five of Ole's drinks soaking in the bottom.

famous Howey Gold Mine. By Christmas, news of the gold discovery leaked out into the national press, and that media exposure resulted in the giant gold rush of 1926. Some 3,000 men from across the continent converged on Red Lake. They fanned out into the bush, travelling on foot, by dog team and even by airplane. The jumping off point was the Canadian National railhead at Hudson. By April 1926, more than 3,500 claims had been recorded.

Another hot spot in the Patricia District gold rush was the Woman Lake area. In March 1926, Tom Powell discovered gold at Woman Lake and soon prospectors swarmed over the territory drained by the Trout Lake River system. Claims were staked on Birch Lake and Narrow Lake. Prospectors chipped free gold from a thick vein which cascaded down a rocky hillside close to the two lakes. A block of Bathurst claims lined the margin of a small lake nearby. A two-mile portage through the bush between the two properties became known as the "golden sidewalk". Anyone walking the trail would dig out free gold with a penknife where the stringers surfaced right beneath their feet.

In the summer of 1926, tremendous activity continued in the Woman Lake area. John E. Hammell, mine promoter, soon founded a company called Northern Aerial Mineral Explorations Company Limited, (usually referred to as NAME). Prospectors were hired for the season for a salary of $150 monthly and flown out to various districts in the interior to prospect for mineral deposits. They were given 10 per cent of the capitalization of any mineral discoveries staked and developed. Captain Harold A. (Doc) Oaks was chief pilot. The company operated 12 aircraft.

Excitement waned for a while because the price of gold was fixed at $20.67 per ounce until 1930. Then, the price jumped to $35 in 1935 and this large increase caused another boom in gold exploration, particularly in the Red Lake area, where approximately 10,000 claims were staked and 100 new mining companies were formed. After a while it seemed that the whole countryside was dotted with tents, and each small camp seethed with activity.

The Roundabout Road to Hudson

My own involvement in this area was precipitated by the stock market crash of October 29, 1929, after which the industrial countries of the world, including Canada, plunged into the Great Depression. At the time, I was employed as a cost bookkeeper and assistant accountant with Dominion Bronze and Iron Limited of Winnipeg, Manitoba, which was a subsidiary of General Bronze Corporation of New York.

In 1932, the plant had about a year's "work in process", including a large contract for bronze showcase window frames and wrought iron ornamental items for the Canadian National Railway's Bessborough Hotel in Saskatoon, Saskatchewan.

Jack Wish outside his bunk house at Project 51.

Despite these contracts, the plant was ordered to close down its operations by the New York office on November 30, 1932. As a result of this layoff, I found myself unemployed and joined the ranks of many thousands of my countrymen who were out of work.

I had gotten married in 1930 and at the time A.G. Duthoit, company treasurer, assured me that I was con-sidered a permanent employee. I felt there was no reason for my not buying a house as it was believed the economic downturn would not last much longer. However, to my surprise, conditions became worse instead of better and I soon found myself out of a job. Although the Canadian government passed a Moratorium Act forbidding any mortgage company to foreclose on homes while the owner was unemployed, I could now see that the Depression would not be of a short duration, so I signed a Quit Deed to our house. Mrs. Wish went back to live with her parents and I left the city for the country to explore a project which I had in mind that might work, at which we might make a living until conditions improved. However, it did not work, so I returned to Winnipeg.

During the Depression, there were so many unemployed and homeless men that the Canadian government passed the Federal Relief Measures Act in 1932 to alleviate the unemployment problem. The act established work camps at strategic locations throughout the country where government work was available and at the same time men were housed and fed. It was more economical than keeping them on municipal relief rolls. The camps were known as "Home for the Single Homeless

Unemployed Men". These camp complexes were given consecutive numbers starting with "Relief Project 1" and so forth. The camps were administered by the militia and the one located about a mile from Winnipeg's Fort Osborne Barracks of Military District No. 10 was known as Relief Project 21.

I managed to get an interview with Lt. Colonel F.B. Eaton of the Army Supply and Transport Branch at the headquarters of Military District No. 10. The interview was a success, and I secured a position as clerk and stenographer for the Colonel on the Relief Staff attached to the Department, at the fabulous sum of twenty cents per day or six dollars per month.

In the spring of 1934, Colonel Eaton asked me if I would consider a chief clerk's position on the Relief Staff of Project 51 in the Army Supply and Transport Branch at Hudson, Ontario. It paid $60 per month, a far better deal than $6 per month. Without delay, I was transferred to the Hudson branch of the Service in early May 1934.

Upon arrival at Hudson, I found Mr. Edgley, the civilian supervisor of the department. Lyle Nash was in charge of the transportation division, which included boats, scows and tractors. Mr. Sutherland was in charge of the food supplies and related functions. I took over the chief clerk's position with a staff of unemployed clerks to handle all the necessary paperwork.

The Hudson train station, soon after the rail line was completed.

A Stranger Comes to Hudson

William (Bill) Stewart resided in Hudson from 1932 to 1937. He was the chief clerk in the Paymaster's Department at Relief Project 51 and was fluent in all Slavic languages as well as German, French and Italian. He was the son of a British diplomat who had been assigned to many different countries, and that's how Bill learned so many languages.

Bill Stewart, left, and golfing companions.

Stewart told everyone that he had joined the Canadian Pacific Railway's Immigration Department and, when the Depression came, lost his job. He then secured a position as chief clerk at the Hudson's Relief Project 51. His story was very plausible and he showed some pictures of himself with some high officials playing golf in London, England. This should have been a clue to the observant. Why was a clerk playing golf with such a distinguished group?

He avoided mixing with any groups at the camp but he eventually became very friendly with me and with Charlie Lucas, the Canadian National Railway's night telegraph operator. Bill Stewart would help me bucksaw wood during the cold winter evenings of 1934-1935 and at the end of the chore, my wife would call us in for some coffee and sandwiches, which was very welcome after coming in from 20 or 30 degree below zero temperatures. We carried on with small talk and played cards. He seemed very interested in the Great War and international politics, and he was always asking me about my views.

During the early part of the winter of 1935, Stewart said that he liked to be outdoors more, so he quit his job as chief clerk in the Paymaster's Department and took a labourer's job at one of the project's camps on Lac Seul. Everybody figured he had a screw loose for giving up an inside job paying $60 per month for a $6 one. These were all clues to a puzzle, but no one connected them.

In the spring of 1935, the G. McLean Grocery Company established a grocery warehouse at Hudson (for the benefit of the mines in the north) and Bill Stewart was hired as its manager, without any grocery store experience. This was another clue that he was not just another ordinary Joe. In 1936, Stewart said that the family was having a reunion in Yugoslavia and, at the same time, there were some property affairs which required his presence. Upon his return from Yugoslavia, he did not elaborate about his trip. Maybe he did not go to Yugoslavia at all. But when the vacation period rolled around again in 1937, he said he was going back to Yugoslavia as he did not see any opportunities in Canada. We were disappointed in him for making these comments and everyone looked down on him when he left.

That was the last we heard of Bill Stewart. Then in 1939 he wired Charlie Lucas from New York, saying that wanted to come to Hudson and see some of his old friends, but "something had come up" and he had to return to Europe.

In 1943, *Look* magazine carried an article entitled "For Once Captain Stewart Guessed Wrong" with a picture of him in an army uniform. It said that Bill Stewart was born in Winnipeg, Manitoba and had eventually become a Captain in Canadian Army Intelligence. It went on to say that he married a Sudetin General's daughter and throughout his stay in Europe and the North African campaigns he was always one step ahead of every move of the German army.

When the Allies were getting ready to establish a southern front, a meeting was held in a Yugoslavian forest where Bill Stewart was carrying out his liaison work with Tito's partisans. The Germans got wind of the meeting and their aircraft strafed the area. Bill had a radio strapped to his back and was in constant communication with headquarters while dodging gunfire from tree to tree, but a piece of stray shrapnel hit and killed him. He was the first Allied soldier killed on the continent (after the fall of France) and his fellow officers were very sad that he was killed, not only because he died but because he possessed so much valuable information.

We will never know why Bill Stewart was in Hudson. We can only speculate that he was either tracking someone down or trying to cover up his tracks. Perhaps his position as the Chief Clerk in the Paymaster's Department was no accident. Where else could he monitor the constant flow of men without creating suspicion?

The base camp of Project 51 had the most professional people in any one place, including bank managers, bank tellers, school principals and teachers, lawyers, accountants, architects and construction engineers. They were not there by choice, but due to unemployment. The base camp's foreman and the chief clerks received a salary of $60 per month, while the rest of the staff and outside workers received 20 cents per day or $6 per month. Each man had a food budget of twenty-one cents per day. The Project received substantial discounts from wholesale grocers and meat packing houses. It also received free buffalo meat from Wainwright National Park in Alberta. These factors helped keep food costs down.

In such a large complex, a considerable number of humorous incidents took place. One of these was a directive received from the Winnipeg Military Headquarters requesting a schedule of transport arrivals at Hudson as well as departures to the outlying camps on Lac Seul. I showed the memorandum to Lyle Nash and he said "tell them that the departures are now and then and the arrivals are if and when".

It was also a point of humour that the Paymaster seemed unable to reconcile the cheques. Since the greater majority of camp personnel were only paid $6 per month, most of it was spent in the canteen and, as a result, cheques were issued to these individuals for only one or two cents. When these men received cheques for such small amounts, they just nailed them to the wall above their bunks instead of cashing them. This resulted in many thousands of cheques outstanding. Hence, a directive was issued that no individual was to be issued a cheque for less than one dollar — anything else was to be paid by cash. This did not create any problem at Hudson base camp, but a special courier had to be dispatched by boat to the outlying camps on Lac Seul in the summer and by dog team in the winter every monthly payday.

The only road in Hudson was a four-mile

stretch running from the centre of the village east to the Keewatin Lumber Mill located on the south shore of Lost Lake and north of the CNR line. Following this road about a mile and a half east of Hudson, where a small bridge crossed a creek, was a large piece of flat land suitable for a base camp. Approximately half a mile east of the creek and bridge there was a short road running north to Cox's Lumber Mill operations, located on the south shore of Lost Lake.

The army engineers decided to build the camp complex in this area. The administrative building contained all the branch offices of the service, except the signal office. It was located in a building on the northeast corner. Across from the administrative building and south of the road the hospital, mess hall, bunkhouses, recreation hall, mechanics' shop and so forth were constructed. The engineers also built a laundry building, complete with a steam laundry and a tailor shop. A water tower was erected about the centre of the main camp complex in the south end.

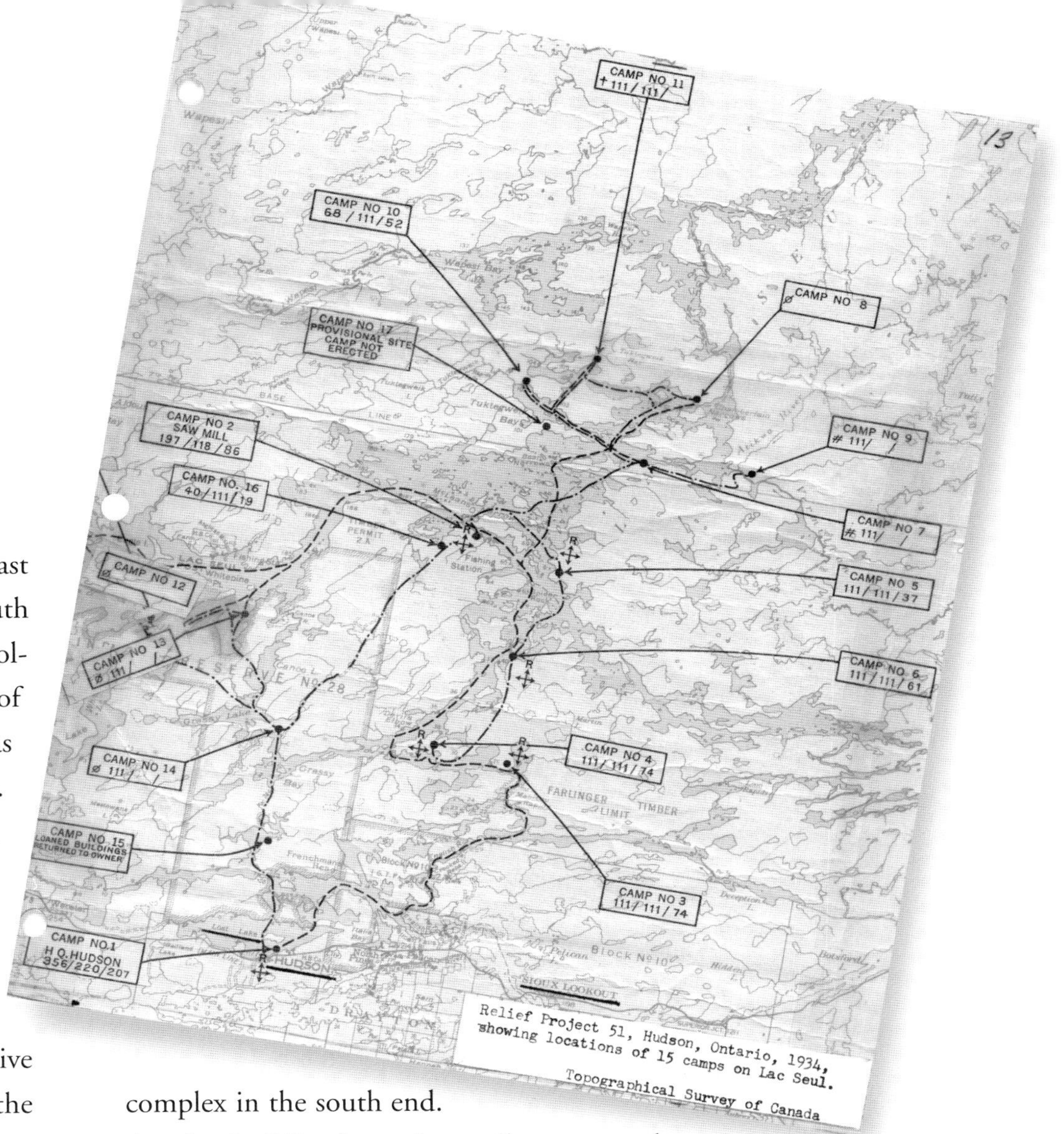

Relief Project 51, Hudson, Ontario, 1934, showing locations of 15 camps on Lac Seul.
Topographical Survey of Canada

Another building housed a small generator plant to supply electricity for lighting. Three large warehouses — one each for food, engineering and ordnance supplies — were built on the Hudson waterfront next to the Canadian National Railway's freight yard.

The engineers then built 15 camps at various locations on the shore of Lac Seul so that

During the Depression, federal government make-work projects employed thousands of men at Northwestern Ontario relief camps.

clearing of the low areas could begin. All the buildings were of a frame construction and covered with tar paper. The project was under the command of Major Vandenberg, and his deputy was Major C.R. Chetwynd. Each branch of the services was headed by a military man, with the exception of the Signal Branch, which was staffed entirely by military personnel. Sergeant Jennings and Corporal Dave Patrick were in charge of the Signal Corps' operations.

There was a certain amount of resentment felt by the local townspeople towards the Hudson camp, as it had a fully staffed hospital. The two doctors were Lerner and Peco. Captain Nichols was the army liaison while Major Cameron of the Medical Corps in Winnipeg made periodic inspections. The camp also had electric lights and running water, which the townspeople envied.

There have been published stories that the executives of the camp had beautiful homes, but I did not see these, outside of Major Vandenberg, who had a rustic type of house built in the Queen Ann style on the shore of Lost Lake near the camp complex. But even this house was not fabulous by any stretch of the imagination. All that one can say is they were not tar paper shacks. Some camp executives were fortunate enough to rent some of the Keewatin Lumber Company's personnel houses, as the mill operations were closed during this period. Housing was at a premium due to the gold rush to Red Lake. Any building with four walls and a roof was considered a house.

The supervisors of each branch, their chief clerks and the camp foreman were allowed to bring their wives to Hudson if they could find houses to rent. It was not until August 1934 that Mrs. Wish was able to join me from Winnipeg. We took over Lieutenant McDonald's house as he was recalled to Winnipeg headquarters. It was comfortable enough in the summer, and was situated on a ridge on Cox Lumber Company's property, surrounded by evergreen trees and overlooking Lost Lake. It had a screened verandah facing the lake and a lean-to kitchen. However, the winters were very cold and I was continually bucksawing wood to keep the fires burning. In the winter, the water in the forty-five gallon steel drum in the kitchen froze over every night. The ice had to be chopped with a hatchet in the morning to get water for brewing coffee and the like.

The camp's occupants included a number of musicians, writers and producers. During the winter of 1934-35, they decided to put on a three-act play for the benefit of base personnel. The play was based on everything relating to the Relief Project. Hudson residents were invited to see this hilarious production free of charge.

Skits lampooning various camp notables were featured. For example, Colonel Candlish, the military supervisor of Army Supply and Transport, always carried a monocle in his coat jacket pocket and when anybody gave him a memo to read, he would squint his right eye and squeeze the monocle into the eye area and say "As you were" in a military command. The monocle would drop out of his squinted eye and hang on the string and he would continue reading with the right eye still squinted. An actor playing the colonel would interrupt proceedings throughout the evening with his sudden "As you weres" and dropped monocles, sending the audience into gales of laughter.

The writers made up a theme song all about the project. The chorus went as follows:

We are the big shots of Hudson
We are the big shots of Hudson
Carnation milk we'll take
Carnation milk we'll take
Because powered milk makes our belly ache.

My employment with the Relief Camp ended in May 1935 when I was successful in securing a position as accountant with the Patricia Transportation Company.

Chapter Two

Freighters, Tugs and Scows

As I recounted in the first chapter, the Patricia District was difficult terrain to live in and to travel over. On the one hand, the many lakes and rivers made transportation a relatively simple matter for prospectors. They could travel by canoe in summer, and by snowshoe or dogteam in winter. But once their gold claims were developed into actual mines, transportation became a much more difficult proposition. Heavy machines and equipment could not be moved back and forth to the mines by canoe or by airplane (air transportation being initially too expensive), so that left two options — by barge in the summer, and by tractor train over the frozen lakes in the winter.

I have devoted this chapter—which details the first days of the Patricia Transportation Company—to summer hauling, which normally involved home-built boats and large scows. The Hudson's Bay Company was the first outfit to haul freight during the summer seasons on Lac Seul. They operated by towing York boats and freight canoes loaded with cargo from Hudson to Ear Falls, and then portaging four times before reaching Red Lake.

The HBC also had a boat called *Ripple* operating on Lac Seul from Hudson to Perch Ripple Portage, with cargo destined for Osnaburgh House located on the northeast end of Lake St. Joseph. Before their cargo reached Lake St. Joseph, it had to be portaged 10 times over a total distance of about 32 miles.

In 1926, Ole Gustafson and Wilfred Wright decided to buy back the Triangle Fish Co. from Arthur Brown and Abe Danto. They farmed out their fish licenses on Lac Seul and went into transportation in a big way. They changed the name to Triangle Transportation Co. and built 100-ton scows to handle freight from Hudson to Goldpines.

There were also many smaller freight operators. Robert Starratt came to Hudson during the gold rush in April 1926, purchased a canoe and outboard motor and began to haul goods and equipment. George and Colin Campbell started canoe

Opposite page: Scows loaded with 500 tons of freight leave for Berens River at the end of the 1934 summer hauling season.

PRIVATE WIRE TELEGRAM

JAMES RICHARDSON & SONS, LIMITED TORONTO, CANADA.

200 . JAR DO RE NWELXK BWDXQMEWBDBYEX AND YOUR PROPOSED TALK
WITH UWDQOW WOYK FOLLOWING WILL DOUBTLESS BE IMPORTANT STOP
THOMPSON AND I SPENT YESTERDAY AFTERNNON RECEIVING VERBAL REPORT
FROM ROFOW AND HE RECOMMENDS WE LEAVE QBDWWDBB EQUIPMENT AND
ORGANIZATION ALONE AND PURCHASE HIGH GRADE USED EQUIPMENT AND ENGAGE
NEW PERSONNEL AS REQUIRED STOP ROFOW WANTS TO COME WITH US IN
CAPACITY HEADING UP NWELXK BWDZQMEWBDBYEX ORGANIZATION AT VERY
REASONABLE SALARY PLUS BONUS ASSUMING WE OBTAIN CONTRACT FROM VEROP
AND DECIDE TO GO INTO NWELXK BWDXQMEWBDBYEX BUSINESS STOP HIS
THOUGHT IS WOK ZDJO NEZK MYXOQ AREA WOULD BE ONLY FIRST OF NUMBER OF
FUTURE DEVELOPMENTS THIS NATURE WHICH ARE NECESSARY IF FUTURE AIR
TRANSPORTATION IN THESE AREAS IS TO PAY STOP . WE DECIDED TO ASK
FOR DETAILED WRITTEN REPORT FROM ROFOW COVERING REQUIRED PERSONNEL
EQUIPMENT ETC AS BASIS FOR COMPUTING COST NWELXK BWDXQMEWBDBYEX

An advisor to James A. Richardson sent him this coded telegram in 1932, regarding the Winnipeg firm's possible entry into the ground transportation business in Northwestern Ontario.

freighting in a small way in 1927 from Goldpines to Red Lake.

The Triangle Transportation Co. contracted to haul Hudson's Bay freight as far as Perch Ripple Portage for their posts on Lake St. Joseph and beyond. They freighted for Charlie Cox who had a mill producing railway ties in Sawmill Bay, after which they were floated into booms and towed to Hudson for loading into railway cars. The Triangle Transportation Co. also hauled a large tonnage of equipment and supplies in their 100-ton scows, which included approximately 3,000 bags of cement for the Ear Falls Dam and hydro generating plant.

In 1931, the owners of the Triangle Transportation Co. decided to expand their freighting operations from Ear Falls to Red Lake. In order to do so, they required more capital. That year they formed a new company called the Patricia Transportation Company.

According to my research (at that time I wasn't yet an employee of the company), the Patricia Transportation Company was incorporated on September 21, 1931 with a capitalization of 50,000 shares of $1 each, of which 30,005 shares were subscribed and fully paid up. The major shareholders were Ole Gustafson, Wilfred Wright and Ernie Wright. The new company purchased from Triangle Transportation Co. the *Archibald* and *Triangle*, along with their scows and other transportation equipment.

The Hammers were Ringing

During the 1932 navigation season, the new company carried out an expansion program. If you showed up in Hudson that summer, you

The **Canvul** *at Hudson in 1935.*

would have noticed a lot of construction activity at the site of the Patricia Transportation Company. The company owners built a new office building, warehouse, icehouse and docks on the Hudson waterfront located on the south shore of Lost Lake. A stiff leg derrick unloaded heavy equipment from railway cars directly onto scows, and handled other equipment and supplies which were too heavy to move by hand.

They also built a bunkhouse for company personnel, as well as a garage to repair tractors, trucks, and other equipment. A diesel DC 25 cycle light plant at the garage site provided power for electric light service at the bunkhouse, garage and the waterfront operations. Electric power was also provided to any company employee who paid for the wiring of his home.

Expansion wasn't limited to Hudson. The company built a bunkhouse, cookery and garage at Goldpines, and a bunkhouse, cookery and icehouse at Ear Falls. A bunkhouse and cookery were erected at Sam's Portage on the Chukuni River, and an office, warehouse and docks sprang up on the Red Lake waterfront. Tractors and sleighs were purchased to carry out winter freighting and additional scows were built for summer freighting.

Ole Gustafson was a Swedish boat builder so he went to work and built four identical boats — *Comet, Patricia, Standard* and *Canvul* — to operate between portages in the Chukuni River waterway, towing scows of freight to Red Lake. He also built the tugboat *Hexagon* and passenger boat *Wapesi* for use on Lac Seul. All this construction activity was expensive and not without an element of risk. But the owners' hard work and optimism paid off when they captured the lucrative Howey Gold Mines' freighting contract from their nearest competitor, Northern Transportation Company, the following year.

By the end of 1933, the frontier competition between the various transportation companies ended and two companies — Patricia Transportation Company and Northern Transportation Company — emerged as victors. They became the two major outfits for transporting equipment and supplies within a range of 200 miles of Hudson. The two companies hauled by boat during the summer and by crawler tractors during the winter.

Passenger Hauling

In 1926 Ken and Archie McDougal, with George Wardrope, decided to enter the passenger transportation business and formed the Red Lake Transportation Company. They purchased a cabin launch, the *Piper*, piloted by Joe Larratt, with Jack Jubenville as engineer. It was the first passenger boat on Lac Seul and was licensed to carry 15 passengers. The passenger rate was $18 for the two-day trip from Hudson to Red Lake. It did not include the cost of the hotel room or meals at Goldpines at the end of the first day's trip.

The popularity of this passenger service was so great that the company purchased *Miss Winnipeg,* a boat licensed to carry 12 passengers. It was piloted by Captain McLean. When the second boat was put into service, the passenger rate was reduced to $12. Either of the boats left Hudson at

7 a.m., and usually arrived at Goldpines at 7 p.m.. The following morning at 6 a.m. the passengers resumed their journey in a canoe propelled by an outboard motor to Slim Collins' (Lower Ear Falls). There, the passengers got into a cart pulled by Collins' two teams of horses and crossed the portage where they boarded the *Pakwash* at Little Canada for the next trip of 32 miles on Lake Pakwash to Snake Falls. They then crossed three portages called Snake, Sam's and Snowshoe. On each portage they crossed by two teams of horses. After crossing the Snowshoe portage the passengers boarded a diesel tug, *Chukuni*, across Gullrock, Keg and Two Island Lakes to their destination at Red Lake. It was not, in other words, a short trip to Red Lake. But it was easier than walking.

During the 1930-1931 winter season, ice conditions were excellent for travelling on Lac Seul. Bob Starratt of the Northern Transportation Company took advantage of these conditions and put in a taxi service between Hudson and Goldpines. Some days he was able to make two or three trips loaded both ways at $15 per passenger for one way.

In order to capture some of this passenger business, the Patricia Transportation Company decided to have a large boat built for this purpose. Ole Gustafson, president of the company, started on the construction of one in 1933. The boat was launched at the opening of navigation in the spring of 1935. It was registered as *Wapesi* (an Ojibway word for swan). The boat was quite a showpiece for Lac Seul — 65 feet long, with a beam of 18 feet and a 205 horsepower Cummins diesel engine. It had two decks, with two lifeboats, and was licensed to carry 24 passengers. It also had a hold below the first deck that

The **Wapesi**, *the Patricia's first passenger boat, was the pride of the company's fleet.*

Company Characters

Dick Roden

In the early spring of 1934, the agent/checker at Ear Falls Portage took sick so the company flew Dick Roden to Goldpines to take over the job until the regular checker returned. Dick had just been hired and was not familiar with the names of the company's executives.

When Dick arrived at Goldpines airbase, which was about a mile from Goldpines and another five miles by water to Ear Falls Portage, nobody identified themselves as being there to pick him up. He saw an elderly man sitting at the end of the dock who looked like a prospector, so Dick approached him and asked if he would take him by boat to Ear Falls. The man agreed.

They were met at Ear Falls by Ole Erickson, foreman, and the rest of his portage loading crew. Erickson showed Dick the bunk that was assigned to him and also the office, which was located in the corner of the bunkhouse. He then returned to his crew and the old man who were discussing navigation problems.

When Dick got oriented with the office space, he came out and saw the old man still carrying on in the group conversation. Dick informed the oldtimer that he did not have any money with him, but if he would send the bill to the Hudson office, he would be paid. The man replied that he would do just that when he returned to Hudson. At this point, the portage crew broke out laughing and introduced Dick to Ole Gustafson, the president of the company. Dick was quite surprised as he never thought the president of a company would be dressed like a prospector, but they all had a good laugh over the incident.

could carry 12 tons of freight. It was the pride of the company's fleet of boats.

However, by the time the *Wapesi* was launched, airplanes were already beginning to emerge as serious competitors. By boat it took a whole day to reach Goldpines, a distance of 110 miles. Upon arrival at Goldpines, the passengers stayed at a hotel overnight. The next morning they entered the Chukuni River waterway, travelled in four boats and walked across four portages before reaching their destination at Red Lake. Furthermore, the boats operating between these portages had no conveniences. By comparison, an airplane ride took one hour and cost only seven dollars more than the boat. Understandably, most passengers soon chose the airplane.

Eventually, the only passengers the *Wapesi* carried were company employees. One of its steady non-paying passengers was Canon Sanderson, who was stationed at the Lac Seul post where he preached the gospel in his native Ojibway language. Whenever he boarded *Wapesi* at either Hudson for Lac Seul post or from La Seul to Hudson, he was issued a complimentary ticket.

One might think that the *Wapesi* was a complete loss as a business venture. But it turned out to be an excellent boat for carrying the weekly perishables to Ear Falls from Hudson. It was also able to pull 100 and 15-ton scows loaded with cargo along the same route.

Tragedy on Lac Seul

As I've mentioned, the navigation route from Hudson to Red Lake covers a course of 135 miles across six lakes, two rivers and four portages.

The Patricia Transportation Company's tugboats operating on Lac Seul were the *Wapesi, Hexagon, Archibald* and *Triangle*. These boats would leave Hudson with large scows and, upon arrival at Ear Falls, the freight would be unloaded onto flat

One of Patricia's scows under construction.

cars and winched across the portage, then unloaded into 15-ton scows. It took several trips to make this transfer of freight and this naturally caused considerable delay along the route.

To eliminate this delay the Chukuni River Portage Company installed a stronger hoist and improved the marine railway so that 15-ton scows arriving on the Lac Seul side could be portaged across without the necessity of unloading and reloading. The new portage system was put into operation at the opening of navigation in 1936.

During the 17 years of navigation on this route (1931-1948), three major accidents occurred. I was present on the scene of the first one, a swamping that took place during a late autumn storm in 1935. We had loaded a 15-ton scow with the portage crew's household effects. Their wives had left a week earlier to escape the bad weather before freeze-up. The same scow also carried a large box containing a Canadian Airways' aircraft engine. The *Wapesi* was the tow boat.

It was a miserable night with freezing rain and snowflurries. Ice was beginning to form on the lake. Captain Wright felt that since the scow was loaded light and the weather was so foul, it wasn't fit for a scowman to remain on the scow while it was being towed, so he asked the scowman to come aboard the *Wapesi*. It was comfortable aboard the *Wapesi* with a warm fire and plenty of food, and we were more or less celebrating the close of the navigation season. Towards midnight, Captain Wright felt that the scow was becoming hard to pull, so he dropped anchor and the deck hands pulled in the tow rope. They found that the scow was swamped, with all its contents washed away. Unfortunately the ice that was forming on the lake cut the sides of the scow, causing the water to leak in.

We remained on the spot until daybreak, when an effort was made to salvage what we could along the shoreline. Various trunks that were still bobbing up and down were salvaged, but beds, bedding and the Canadian Airways' engine were not recovered.

By this time we were overdue at Hudson, so the company management and the wives of the crew were very concerned. No plane could be sent up in the snowstorm to try to locate us. Towards evening we finally arrived and docked. Everyone was glad to see that we were none the worse for our experience. However, the greeting party lost their enthusiasm when they realized all their possessions were gone, with the exception of the clothes in the wet trunks.

The second accident occurred on Lake Pakwash between Ear Falls and Snake Falls during the summer of 1936. Lake Pakwash is a shallow lake plagued by sudden storms. It had a record of claiming a life every season.

One day Captain Carl Robinson pulled out of Ear Falls with his *Canvul* for Snake Falls. He was towing six 15-ton scows, three of which were loaded with dynamite. Halfway across, he got caught up in a sudden storm. All his efforts to reach the nearest shore were unsuccessful. The *Canvul's* propeller was out of the water more than in, and the wind was blowing them towards a rock island. Robinson was afraid that if they hit that, the impact might cause the dynamite to explode and the boat would be blown sky-high. Fortunately, this did not happen and a few days later the Howey Gold Mines' dynamite crew came by. These experienced dynamiters towed the scows loaded with explosive to an isolated spot and blew the load up. Forty-five tons of dynamite made quite a bang!

The third accident occurred during the summer of 1937 on Lac Seul, when the *Wapesi* was

Trying to beat the freeze-up on the Chukuni River in 1936.

on its way to Ear Falls. An electrical storm came up and a bolt of lightning struck Eddie Snowball, the scowman, who fell overboard. By the time *Wapesi* could stop and return to the spot where he went overboard, he was nowhere in sight. His body was recovered three days later. This is the only loss of life that the Patricia Transportation Company experienced throughout 22 years of navigation operations in the area (1931-1953).

Summer Hauling on the Root River

Despite the name, the terminus of this navigation route was not Root River but at Doghole Bay, which is located at the northern end of Lake St. Joseph, where the Albany River embarks on its long journey to James Bay.

After freight was unloaded at Doghole Bay into the respective mines' warehouses, it was hauled by truck another 27 miles to the mine properties over an all weather gravel road. The first year of operations Hennessy Trucking Company carried out the job, followed by Haverluck and Koval up to 1947, when the Patricia Transportation Company took over. When the route opened for navigation in 1935, the Patricia Transportation Company secured an exclusive contract for hauling of Central Patricia Gold Mines' freight. They also had the Hudson's Bay Company freighting contract.

At the head of the Root River where the three-and-a-half mile portage led onto a bay at Lake St. Joseph, a gas Whitcome "Dinkey" locomotive was introduced to pull freight on regular flat railway cars across a standard gauge railway. In later years, the gas-operated Dinkey was replaced by one operated by diesel power. Two stiff leg frames were installed for loading and unloading at either end of the portage. The first year the Dinkey was operated by Bill La Rouse and during the intervening years by Louis Lacoste and Paddy Cochrane.

Opposite page: After crossing three marine portages, the* Root River Scow *navigates a narrow waterway.

Patricia's open air cookery at Nattaway Portage in 1935.

Wood fuel was used for all the steam boilers on the three portages, although coal was tried for one season at Nattaway. Although Patricia Transportation Company's river crew was based at Nattaway Portage in 1936, it did not provide any facilities. Captain Shorty Livingston slept on the boat *Flatty*. He was short enough so that he could lie crossways on the floor between the engine and the steering wheel, while Sammy Learmonth, the engineer, slept on a bench in between the propeller shaft and the side of the boat, behind the engine. There was no cookery at Nattaway so the crew rigged up a pole frame and covered it up with one of the Hudson scow tarpaulins.

Sammy Learmonth relates that on one occasion at Flour Portage, the steel pin holding the cable to the portage car came out and the loaded scow rolled back and struck another scow loaded with 12 tons of dynamite waiting its turn in the slip to be portaged. Needless to say, the crew went through a very tense moment seeing the scow rolling backwards towards the loaded dynamite scow. Fortunately when they collided, there was no explosion.

In 1935, Ernie Wright of Patricia was put in charge of establishing the company's navigation route and providing the necessary facilities for its operation. He used the motor launch *Arrow* on the Root River waterways portage system. The motor launch *Marie* was used for towing scows from Root Bay on Lake St. Joseph to Doghole Bay. There, Harry Everett experienced a near-disaster of his own.

During the day, a shipment of dynamite arrived by boat at Doghole Bay and three trucks were loaded to be ready to leave the next morning. A loaded truck was parked on each side of the office building and one in front. A severe electrical storm started and Everett knew that if the lightning struck any of the three trucks, the entire Doghole Bay waterfront would blow up. He considered moving the trucks but was afraid that if he started any one of the trucks, it might draw lightning to it. So, he sweated out the storm, and said later that it was the longest night of his life.

From that point on, whenever trucks were loaded with shipments of dynamite they were moved up the road from Doghole Bay and parked overnight a safe distance from town.

Throughout the years of navigation on this route, the Patricia Transportation Company experienced two cargo losses as a result of severe storms on Lake St. Joseph. One occurred in September, 1939 when Captain Stewart Vincent pulled out of Root Bay with *Lac Joe* towing the usual two 45-ton scows loaded with a carload of sugar (140,000 pounds or 70 tons) and about another 20 tons of canned goods bound for Central Patricia Gold Mines at Doghole Bay. This stock was supposed to last throughout the winter.

Flour Portage on the Root River.

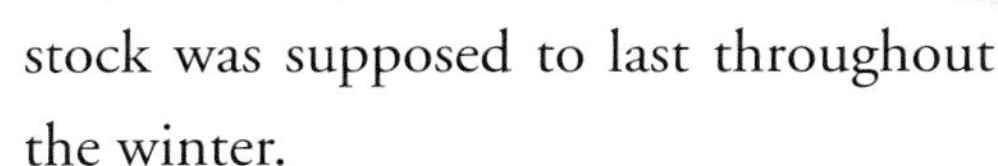

About halfway to his destination, a severe storm developed and Captain Vincent did not have a chance to reach shore for shelter. The scow swamped in the waves and 25 tons of sugar dissolved and the sacks just floated away. The balance of the cargo in the scow was also water damaged. When the storm abated, Vincent proceeded to Doghole Bay with one loaded scow as well as the swamped scow in tow. Upon arrival at Doghole Bay, it was discovered that the paper cartons containing the canned goods had fallen apart and all the labels on the cans had come off, so no one could tell what

Doghole Bay

Doghole Bay is in the Patricia District at the north end of Lake St. Joseph, the source of the Albany River. It is 20 miles south of Pickle Lake, 21 miles from Central Patricia Gold Mines and 27 miles from Pickle Crow Gold Mines. It was the terminus of the 186-mile Root River Route from Hudson. Ontario Hydro's Rat Rapids generating station, located on the Lake St. Joseph-Albany River outlet, was about one-and-a-half miles by water and one mile by road from Doghole Bay.

The Hudson's Bay Company's Osnaburgh House, which was built in 1796 during the fur trade era, was approximately seven miles south of Doghole Bay. The post was closed in 1810 and reopened in 1841.

There was an approximately 130-foot long dock running along the lakeshore with a small portion running out from the main structure for docking of smaller boats. The forestry pump was located at the end of it all summer, in case of fire. At the end of the main dock was a manually-operated wooden derrick used for unloading heavy and bulky freight. At the other end were two North Star Oil Company storage tanks, one for fuel oil and the other for gasoline. Near these tanks was a sheet metal warehouse building where oil products were stored.

Murdy Hardy was the North Star Oil Company's agent as well as Starratt Airways and Transportation Company's. He also did the billing of freight when Koval Bros. had the trucking contract for hauling freight from Doghole Bay to the mines. He was dubbed the mayor of Doghole Bay.

Central Patricia and Pickle Crow Gold Mines had a warehouse each for storing of their equipment and supplies upon arrival of the loaded freight scows from Hudson before it could be trucked to the respective mines' properties. Tarpaulins were provided to protect the heavy equipment and the overflow of other freight stored outside for which there was inadequate storage space in the warehouses.

Hennessy started trucking to the mines upon the completion of an all-weather gravel road from Doghole Bay to the area in 1935. Koval Bros. took over the trucking in 1944 and Patricia Transportation took over in 1946 and continued until the cessation of navigation to Doghole Bay at the end of the 1953 fall season.

Hennessy had the facilities for their road building crews and truckers, such as bunkhouses, cookery and a garage for the repairing of their equipment. The two transportation companies also had their own bunkhouses and cookeries for their personnel.

Doghole Bay was a hive of activity for 19 years (1935 through 1953).

Opposite page: Indian "bull gangs" at Doghole Bay unload freight following a 225-mile run.

Swamped scows.

they contained. Stan Bosigar, the company agent, telephoned Central Patricia Gold Mines' office and ask them to relay the message to the Hudson office informing them of the accident.

Upon receipt of this message, the cargo insurance people in Winnipeg were contacted and they sent an insurance claim officer to examine the loss. Time was of the essence, as the mine would have to reorder the lost merchandise and have it delivered within the next four weeks before the close of navigation due to freeze-up.

After receiving the urgent message, the insurance representative left Winnipeg that evening, arriving at Hudson at midnight on the Canadian National transcontinental passenger train heading east to Toronto. In the morning, he and I boarded a Canadian Airways' plane for Doghole Bay. On arrival at Doghole Bay, the insurance representative surveyed the damage, which was a mess — cans, paper cartons and other freight strewn all over the bottom of the scow. He ordered that all the canned goods be given away to the local Indians. Of course, he had no evidence of the sugar loss as all the bags floated away when the sugar dissolved during the storm. However, the sugar loss was substantiated by company waybills and the claim was put through.

The other accident occurred in the middle of summer in the mid-1940's when a scow was swamped on Lake St. Joseph with a shipment of tobacco and cigarettes worth about $35,000. The scow was also loaded with feed oats, powdered soap, dry goods and a Mercury car. U.M. (Slim) Dart took care of this accident by flying out from Hudson with an insurance adjuster.

The car's back seat got wet but the front was okay, so the adjuster offered the consignee $750 in settlement. The cigarettes were only slightly damaged by the water but, of course, were not saleable, so the crews at Doghole Bay had free cigarettes for weeks afterwards.

In modern life we take perishable foods like dairy products and eggs and meat for granted. But hard-working men in remote mining camps appreciate good food, in abundance, and such foods were hard to come by until the Patricia Transportation Company initiated a perishable freight delivery service from Hudson to Red Lake. This service continued until 1948, when an all-weather highway was constructed from Vermilion Bay to Red Lake.

This is how the system worked: the Canadian National Railway freight train arrived at Hudson from Winnipeg every Thursday afternoon and parked a refrigerator car on the waterfront siding loaded with perishables. The company crew iced the bottom of the scow and the perishable freight was loaded on top of that layer of ice. When the loading was completed, a tarpaulin covered the merchandise and another layer of ice was spread over the top.

Once the scow was loaded, one of the tugboats would take it in tow for the next 110 miles to Ear Falls, arriving there early Friday morning. As soon as the barge arrived at Ear Falls, the loading crew iced the bottom of a 15-ton scow on the English River side of the portage, then unloaded

A refrigerator scow at Red Lake in 1937.

The marine railway at Sam's Portage in 1936.

the perishable freight from the Hudson scow onto the marine railway flat car and winched it across the portage where the perishables were reloaded into the iced scow. It was then covered with a tarpaulin and a blanket of ice was layered over it.

The Chukuni River boats then took the load over the next 65-mile haul. Then the load was portaged over the Snake, Sam's and Snowshoe Portages. It was then towed to Red Lake, arriving there sometime Saturday morning.

This method of handling perishable freight was carried out for 10 years. During this period of time there was some competition with Starratt's as to whose reefer would arrive first at Red Lake. Patricia's crew won out most of the time due to Captain Wilfred Wright's ability to travel during the night. He had fished on Lac Seul all his life and he knew its waters like the back of his hand.

In 1946, the company retired their wooden boats from service and replaced them with steel boats which could easily be portaged in the same manner as the 15-ton scows. The new steel boats displaced 10 men, i.e. six of the river crew and four in the cookeries (which were eliminated at Ear Falls and Sam's). The tugboat *Patricia* was brought back to

Hudson and used for "warping" scows, that is, switching them around as required and lining them up for their departure. The handling of the reefer service by the steel boats from Hudson to Red Lake lasted for two years until the highway was constructed in 1948.

The Boys in the Office

So much for summer hauling. When fall arrived, the Red Lake mining companies and merchants placed their orders with packing houses for their entire winter's supply and the merchandise had to arrive in Red Lake before freeze-up. The responsibility was entirely upon the transportation companies. If the perishable freight was shipped too soon, it would spoil en route. If it was shipped too late, it would not arrive at its destination in time before freeze-up. Of course, the weather had a mind of its own and predicting the most opportune time for

Company Characters

Charlie Wilson

Charlie Wilson was the managing director of the Patricia Transportation Company, and he was famous for being tight with a nickel.

When it came time to register Ole Gustafson's four lookalike boats - *Comet, Patricia, Standard and Vulcan* - Charlie Wilson ordered the brass letters for the names, and sent the applications to the registrar in Kenora for licensing. The names were all accepted, with the exception of *Vulcan*, for there was already a boat registered under that name.

Charlie Wilson couldn't see the point of throwing away the brass letters for *Vulcan*. So he just mixed up the letters and named the boat *Canvul.*

making this late fall meat run was an art form in itself.

Charlie Wilson, the managing director of the Patricia Transportation Company, had an uncanny ability for forecasting the weather. He and Don McLennan, the firm's accountant, lived across the street from one another and about an eighth of a mile from the CNR tracks. Instead of walking to work through town to the office, they took a shortcut to the track and walked along it to the waterfront office. Charlie liked being outdoors and he had a good feel for nature.

Some years, at the beginning of October, it would be cold with snowflurries during the day and freezing temperatures during the night. While they walked to work Mac would say, "Charlie, don't you think I should wire the packing houses in Winnipeg and have the perishables released for shipment to Red Lake, as the way the weather has been this past week we might have an early freeze-up?" Charlie would shake his head. "Naw, Mac, it's going to warm up." Sure enough, in a few days the weather would change and it would be quite warm for a week or so.

On other occasions the weather would be steady and beautiful. The two men would walk to work on a warm day, with the sun beating down and golden leaves blowing across the rails, and Charlie would say: "Mac, send the wires to the packing houses and have the perishables released." McLennan would protest: "But Charlie, it's been so warm all this month. We'll spoil the shipment for sure." And Charlie would reply: "Mac, it is going to turn cold on the day the shipment arrives on the train."

And he was always right. Charlie Wilson knew the weather. And our freighting competitors watched his every move. When Charlie wired for the perishable food, they all jumped to do the same.

Opposite page: The water-front at Hudson.

NAR
Caterpillar
Diesel
CATERPILLAR

Chapter Three

The Challenges of Fire and Ice

As I've explained, a gold discovery in late 1925 created the famous Red Lake Gold Rush. Men from all parts of Canada came by train to Hudson, Ontario hoping to find their pot of gold. It was bitterly cold that winter and the only way to get to Red Lake, a distance of approximately 135 miles, was either by dog team or on foot (many hundreds of men walked).

It soon became necessary, even in winter, to move larger loads to successful mine sites. People tried all kinds of methods for hauling freight on the ice. The Great West Transportation Co. had several Model T Ford trucks with snow tire chains on their rear wheels for hauling across Lac Seul from Hudson to Goldpines. The Red Lake Transportation Co. used a REO Speedwagon truck hauling two-ton loads of freight from Hudson to Goldpines on Lac Seul.

The Northern Transportation Co. got a contract to haul 40 tons of freight from Hudson to the Hudson's Bay post at Woman Lake. They rented 16 horses and eight sleighs from Carter Hall Co. of Winnipeg to carry out this work. Many of the horses perished from the tough travelling conditions in subzero weather, so the horse teams were replaced by Fordson farm tractors hauling two sleighs of customer freight, including their own supplies, such as fuel for the tractors and so on.

These early machines were not suited for snow conditions on Lac Seul as their pulling power was limited. Furthermore, they were very unstable and, when going up a hill, had a tendency to rear over backwards and pin the driver. On one of these occasions a tractor tipped over on the English River and killed the driver. A similar accident occurred on Dead Man's Hill near Red Lake, hence the present name. It soon became evident that permanent use of these machines was an impossible proposition. They were not only dangerous and underpowered, but when they encountered problems such as slush, tree stumps and the like, the trip was often a complete loss.

Opposite page: This Caterpillar tractor hauled 37 tons of freight through snow that was at times 4 1/2 feet deep.

Mel Parker driving the reliable Cletrac tractor.

Opposite page: C.T. Wilson (left) and Ole Gustafson in 1936.

When the Triangle Fish Company was reorganized and incorporated in 1931 as the Patricia Transportation Company Limited, it replaced its Fordson farm tractors with gasoline crawler tractors, 25's, 30's, 35's and 55's. The Cletrac 55 was a powerful machine that could haul heavy loads, but by the same token it was a gas guzzler. For example, it required a full sleigh load of 45-gallon gasoline drums to make the 135 mile trip to Red Lake. This took up a sleigh which could have been used for paying merchandise. So it was a relief when diesel tractors replaced the gas-powered machines, and only a few drums of fuel oil were required for a two-tractor swing.

When the Cleveland Tractor Company came out with their diesel Cletrac Crawler tractors in 1936, they sent a sales representative to promote the sale of their new machines for use in the north country. The salesman called on the Patricia Transportation Company and a meeting was held at the company's Hudson office with Ole Gustafson, Wilfred and Ernie Wright, C.T. Wilson and myself, at which time the salesman extolled the virtues of their diesel tractors. He said that Admiral Byrd used them in his Arctic Expedition and the tractors operated very satisfactorily in severe weather. Charlie Wilson, managing director, with his tongue slightly in cheek, said, "Ah, but did he encounter slush up there?"

We were satisfied with their gasoline Cletracs, so the company went ahead and ordered 13 of the new diesel 40's for the hauling of freight and one 35 which was used for trail breaking and scouting. It was dubbed the "puddle jumper".

We also ordered a couple of RD6 diesel tractors from the Caterpillar Company to see how well they would perform in the Hudson area. One might think that hauling freight with a diesel tractor is a simple matter of putting the tractor in gear and heading off. This is certainly not the case. In the north country a tractor train must cross many portages, combined with many lakes and rivers — each one presenting varying conditions of weak ice and deep slush.

The total dead weight of a crawler tractor, one sleigh and cargo is approximately 20 tons, covering an area of 240 square feet. It took a special man to cope with rough trails, bucking 15-ton sleighloads of freight, rough portage trails and a hostile environment. Freighting crews were away from their homes and families from three to four months at a time. They were constantly on the move, working eight hour shifts without any days off for Sundays or holidays. They worked outside, where the temperature was almost always below zero, and sometimes plunged to 40 below. There was also the problem of snowstorms and blizzards. A man who couldn't "cut the mustard" wouldn't last through a shift.

A caboose in severe slush conditions at Manitoba Point on Lac Seul.

After several years of experimenting with different tractors, the Patricia Transportation Company discovered the best equipment was a 40 HP diesel Cletrac crawler weighing approximately seven tons, hooked to a sleigh with four inch steel runners and rack weighing a total of two tons. This rig pulled a caboose of frame construction with tongue-and-groove lumber on the outside and inside walls. The roof was covered with heavy duty roofing paper. The caboose was 18 feet long, eight feet wide and seven and a half feet tall. There were six bunks — three bunks on each side of the wall — and a three foot aisle in between. The caboose provided a precious place where hardworking crews could seek shelter from the bitter cold.

One item which often caused a little excitement was the stove. It had a steel lip welded around the top and this was intended to prevent the cooking utensils from flying off, but when the going got rough on a bumpy trail, the pots often went flying anyway. The cook had to maintain a balancing act while he was at work at the stove, and it was not unusual for the caboose to hit a rough spot and the frying pan and pails would fly off the stove, upsetting the whole mess. This would be followed by the cook swearing profusely. On such occasions it was best not to say anything and keep out of the cook's way or you might end up with a frying pan slammed over your head.

The rest of the caboose was filled with cupboards, bins and a table where the tractor crews could sit and have their meals. On the outside of the caboose, next to the door, a cupboard was built onto the caboose with shelving and a door where meat products were stored. No fresh vegetables were used as it was impractical to prepare them when moving over the rough trails. The balance of perishable goods were kept in cartons under the bunks to prevent them from freezing. All utensils used were tin, as stoneware would break and

enamelware would chip due to the rough travelling conditions. A coal oil lantern hung from the ceiling in the living area. Coleman gas lanterns were not supposed to be used due to the danger of an explosion should a caboose overturn. But some crews disregarded the rules and used them anyway.

The drivers and brakemen wore clothing designed for the harshest conditions. Some wore two suits of woolen underwear and two pairs of woolen socks. Heavy rubber boots tended to freeze the feet, and felt boots also froze when they got wet. So footwear usually consisted of two woolen pairs of socks tucked into a thin felt boot, sometimes called a German sock, over which a sealskin mukluk was worn. The sealskin mukluks were soft and pliable, would not freeze and were also waterproof. The men also wore parkas, fur caps and heavy mittens. With all this clothing their movements were somewhat restricted, but it was the only way to keep warm while sitting on a tractor seat in subzero weather.

It was windy and bitterly cold on the open seat of the tractor, but it was impossible to surround the driver with a cab because if a tractor broke through the ice, the driver would have no chance of jumping off or extricating himself from the cab. In other words, a cabin would keep a driver warm but it would also spell his death if there was an accident. Even without the cabs, some drivers didn't manage to jump clear when the tractor broke through the ice.

Each swing of three tractors had a crew of nine men, consisting of a swing boss — who was also one of the tractor drivers — and five other drivers, two brakemen, and a caboose cook. The swing boss, apart from being a tractor driver, was also a capable mechanic. If any problems developed when

Arvid Anderson and Mike Matty stand in front of the caboose cook and the "brakie", somewhere on the trail to the Berens River Mine.

Company Characters

Don McLennan

Don McLennan, the accountant/secretary of the company, was a workaholic who spent most of his time at the office. In the mornings, while reading the *Winnipeg Free Press,* he would peek over the top of it to see if Dave, Slim and I were busy at our desks.

When we wanted to be alone for a little while to goof off, Dave figured out a way how we could get Mac out of the office for a spell. During his lunch hour, Dave would buy a girlie magazine and bring it back to the office and place it in full view on the top of his desk and proceed with his work. Mac would pass by Dave's desk several times and glance at the magazine. Then he would say, "Dave, can I look at it for a while?" Dave would reply, "Sure Mac, help yourself — I don't need it until I go back to the hotel after work." Mac would pick it up, go to the upstairs office and — if it was a particularly good issue — we would not see Mac for about an hour or so.

he was off shift, he would be rousted out of bed to take care of the situation. It was all part of the job.

During the winter hauling season, the tractor swings operated 24 hours per day without any holidays. The train paused only to refuel and to change shifts. The shifts were eight hours, unless the weather was so severe that the men couldn't take the exposure and it was necessary to cut their shifts down to six hours. While the crew coming off shift refueled the tractors, the crew coming on duty ate their meals before taking over. After refueling the tractors, the crew would come into the caboose and have their meal when the swing was already moving. It was an art to balance the food on the plates when the swing was moving over the rough trails. After finishing their meal they would remove their outer garments and crawl into their respective eiderdowns in the bunks for rest and some welcome sleep.

At first, one tractor was used to pull a train. However, from experience it was found that it was

Drilling holes for pine tree trail markers.

Previous Page: "Doubling-up" near Lac Seul in 1940.

better to use two tractors in a swing, as an additional three sleighs of paying freight could be hauled. Furthermore, if one tractor went through the ice, there was another to help out or to proceed to the destination with a lower payload. At that point another crew would come by and salvage the sunken tractor and deliver the balance of loaded sleighs if they too did not go through the ice.

It was also an advantage to have two tractors in a swing for "doubling-up" while crossing portages. This enabled the swing to make better time in going over them. If slush conditions were encountered, in most cases the crews had to unload the freight off the sleighs, chop the sleighs out of the frozen slush, winch the sleighs by cable to a dry spot and reload them before they could proceed any further. Sometimes they might encounter two or three different areas of slush conditions. If a tractor fell through the ice, it had to be dried out and completely overhauled.

Before the first tractor train headed out, however, a crew would go out and prepare the trail, They would provision a caboose with food supplies, and load a sleigh with supplies, drills, and diesel fuel oil for the IHC 20 HP crawler tractor. The crew consisted of four men–a tractor driver, a cook and two labourers. The men cut a good supply of small evergreen trees and piled them on the sleighs. Crossing the lakes, they tested the ice with a special two-inch auger which was equipped with a handle long enough to drill a hole in the ice in a standing position. In this test hole they inserted an evergreen tree. The purpose of testing the ice was to determine whether it was thick enough to support heavy loads. The evergreen trees placed in those holes about 100 yards apart marked the trail on the lake or rivers, and the tractor driver used the trees as a guide for the route. Without these trees, the tractor driver might stray onto thin ice or lose his way in a blizzard.

After testing the ice and marking the trail with evergreen trees, the crew had to tamp down the muskeg areas so they would freeze better and support the heavy loads. The crews also had to see that the approaches to portages were in good condition and, if not, they had to make the necessary improvements and clear the trails of any windfalls that occurred during the summer.

If it had been a mild fall and the muskeg areas hadn't frozen solid enough, they were tamped down by a light International 20 HP tractor equipped with a long pad on the tracks. Some years it was necessary to make the ice stronger on the rivers by putting down long logs and then shovelling slush over them so it would freeze and make the ice strong enough for the tractor swings to cross over.

It took the crews about 10 days to carry out this work, which could only be done during daylight hours. Sometimes it took a little longer, depending on the conditions they encountered on their journey to the mine. Upon reaching the mine, they would send word back that the ice was safe and the route thoroughly marked for tractor swings to proceed with the sleigh loads of freight.

Sometimes a tractor would break through the ice at approaches to portages, but typically the water was not very deep and a tractor could be salvaged and they would carry on again. Allan Sandin was usually given the job of trail breaking at the beginning of a winter freighting season in Berens River. Sandin related that when he was a swing boss

on one of the tractor swings on his way to the mine during the winter season of 1940-1941, he came very close to meeting his maker.

The accident occurred in McIntosh Bay on Deer Lake near Oskar Lindokken's trading post, about 30 miles from the mine property. Although the regular trail was followed, a crack must have developed under the snow over time and the tractor went through the ice with Sandin still on it. Fortunately, he bobbed up in the open hole and the crew pulled him out to safety, soaking wet but none the worse for his experience. He was rushed to the caboose for a change of dry clothes and a good slug of antifreeze (Hudson's Bay brandy).

After the accident, the swing carried on to the mine property with the remaining two tractors and all their loads, leaving a tractor sitting on the bottom of the lake in 165 feet of water. Sandin said that he and Ernie Wright returned to the scene of the accident three days later to carry out the salvage operation. It took them three days with chain blocks to raise the tractor, put it on skids and tow it to shore, where they removed as much mud as possible with the help of a big fire nearby. It was then loaded on a sleigh and hauled to Berens River for a complete overhaul. The accident was costly and time consuming, but at least Sandin escaped with his life.

Many tractors went through the ice. On January 1, 1934, Ernie Wright was driving a tractor with Dale Wilson as his brakeman. They went through the ice at Manitoba Point on Lac Seul with several sleigh loads of Howey Gold Mines freight. Both of the men were rescued when they bobbed out of the hole. They had the caboose on the swing and after they dried their clothes, they walked to Goldpines for help. The tractor and some freight was salvaged by using the services of Joe Bernier, a local diver.

Ice conditions were usually very poor on the route to Red Lake and Wilfred Wright went through the ice on several occasions. One time he put a tractor through the ice on Woman Lake (now known as Confederation Lake) on his way from Goldpines to the Argosy Gold Mines at Casummit Lake. The tractor went down in 80 feet of water and Wilfred was fortunate enough to bob up in the hole and was pulled out. The crew said it was so cold that by the time they got him to the caboose, his clothes were frozen solid so they just cut them off and had him consume a large quantity of Hudson's Bay brandy. He suffered a very bad cold, but otherwise he was none the worse for wear. Another crew spent a week trying to salvage the tractor. When they got the machine hoisted on a tripod, it was put on skids and hauled to the Goldpines garage for a complete overhaul. There was no damage to the tractor in spite of its going down so deep in the water.

Even when they were off shift and sleeping, the fear of drowning persisted in every crewman's subconscious mind. On one occasion, Ole Erickson was off shift and sleeping in his bunk. The caboose was going over a portage trail and when it went over a windfall it gave a great lurch. Ole Erickson

Opposite page: Driver navigates through dangerous slush conditions on Lac Seul in 1934.

Trucks Versus Tractors

Although mine operators carefully estimated their perishables requirement when placing their orders in the fall of the year, certain items would be underestimated. So when the winter freighting season arrived, they often ordered additional items to carry them through until spring. The company provided a heated caboose for the hauling of certain fragile perishables such as eggs, while carcasses of meat and the like could be hauled on an open sleigh as it did no harm to freeze them.

Some trucking companies from Winnipeg felt that there was no reason why truck hauling to Red Lake could not be successful, and they suggested that they could do it at a considerably lower rate than the local tractor hauling companies. In 1936, the mine operators and merchants of Red Lake advised both freighting companies in Hudson (Patricia Transportation and Starratt's) that they had made other arrangements to handle their winter freighting needs. They told us that they were going with a Winnipeg trucking firm, and that their freight was going to be delivered at a much lower rate.

Despite this news, both companies went ahead and got their winter freighting equipment ready. We knew that trucks could only be used in certain seasons when ice conditions were favourable on Lac Seul. This occurred rarely. The winter of 1931 was favourable for trucking, as was the winter of 1938. But most of the time the lake was far too slushy for trucks. To make good ice for truck traffic it was necessary to have a special year with an early freeze-up, lots of cold weather, and a light snowfall.

The winter of 1936-1937 was not one of those years, and there was a lot of deep slush on Lac Seul. So without knowing this, the trucks from Winnipeg arrived at Hudson, loaded their freight, and departed for Red Lake at daybreak the following day. When I arrived at the office at 8 a.m. that morning, I could see that one of the trucks was stuck about one half mile from the shore.

From the office window, we watched the truck crew trying to dig it from the slush. The other truck came up to help, but it got stuck too. The crews of both trucks were now wading around in the slush, trying to extricate their vehicles with shovels. A little further past them was an island. Ole Gustafson said "by the rate they are travelling I figure they'll reach the island by lunch time. And by nightfall they will be back here on shore."

Ole was right. The two trucking crews eventually wrestled their vehicles back onto shore, unloaded their freight and returned to Winnipeg, no doubt resolving that they should never have attempted to challenge local transportation companies who had been operating in the area for many years.

With hat in hand, so to speak, the mines and other customers asked us if we would reconsider and haul their freight. We informed them that since they had already advised us early in the season that they had made other arrangements, our equipment was not prepared and we would have to charge them extra. We did not in fact charge them extra, but we wanted to make a point. And this was the last time anyone tried to compete against the two well established companies (Patricia and Starratt's) for winter freighting in the area.

Diver Joe Bernier of Hudson about to retrieve a sunken tractor.

woke up with a start, jumped out of his bunk, flung the door of the caboose open and jumped out into the cold in just his underwear and stocking feet. When he came to his senses, he started to run after the caboose. The tractor swing was only travelling at about four miles an hour, but Ole said later that it was difficult trying to run in the snow in his underwear and stocking feet.

There were many accidents involving tractors going through the ice. Even with all the accidents, the Patricia Transportation Company was very fortunate in not losing a single man through drowning in their 25 years of winter freighting operation in Northwestern Ontario.

The Patricia Transportation Company hauled throughout the north, including freighting supplies across Lake Winnipeg, from the Berens River Mines to the railhead at Hodgson. Fifty miles was over a bush trail and approximately 30 miles was across the lake. The first year the hauling started with four trucks pulling two sleigh loads each with a snowplow on the lead truck. Later, TD 14 tractors were used. These tractors had to be equipped with special ice grousers in order to travel over glare ice.

Lake Winnipeg ice shifted a lot and created wide cracks and ice ridges. It was not unusual for a tractor to suddenly encounter a wide gap in the ice. The gaps ran each way for miles and there was no way that the tractors could go around them. The crews handled such emergencies by carrying 12"x12" heavy timbers with them. They laid these long timbers across the narrowest crack, and slushed them over with water and snow so it would freeze. After it was sufficiently frozen, the tractor would cross the bridge first and then one end of a 20-foot steel cable would be hooked to its drawbar and the other end to the sleigh's cross arms. The sleighs would then be brought across singly to the other side. When all the sleighs were hauled across, they were re-assembled and the swing proceeded on its way again. The timbers were picked up and loaded back on the equipment sleigh for use again whenever another crack was encountered.

During the first winter of operations on Lake Winnipeg, a caboose was upset twice. The caboose wasn't attached to the sleigh, so that if the sleigh went through the ice, the sleigh would sink and the caboose would float long enough to enable the crew to escape drowning. Sometimes the driver would forget he was pulling a caboose, especially in trying to make a small hill. If there was a sharp turn at the bottom of the hill, the sleigh would swing out and hit a hard drift, turning the caboose on its side.

This happened one time after a swing left Hodgson on the way to the lake. We stopped our truck and waited for the caboose to appear so we could have coffee before going on the lake. The truck hauling the caboose finally showed up with its two loaded sleighs and two sets of sleigh bobs, but no caboose. The driver got out of his cab and we asked him where the caboose was. You should have seen his reaction when he saw that his caboose was missing. We unhooked the sleighs and went back about two miles when we saw the cook standing in the middle of the road, his face splattered with what looked like blood.

The first thing the cook said was, "I am walking back to Hodgson as I have had enough of

Ice heaves like this one on Lake Winnipeg presented a barrier that often extended for miles.

Despite precautions, dangerous "jack-knifes" sometimes occurred.

this job." We assured him that we would clean up his caboose, which was a mess — wood ashes mixed with syrup, jam, rice, beans and the like. Cleaning it was quite a task, but more importantly, we convinced the cook to stay on and finish the season. What he thought was blood on his face was, in fact, beet juice. He had some beets in a quart jar on a shelf, and when the caboose tipped, the jar hit the opposite wall, shattered and splashed juice all over his face. While cleaning up the mess, another truck came by with the empty sleigh bobs. It was some job getting the caboose back on the sleigh as it was necessary to right it and get the bobs under, one set at a time.

When accidents occurred, rigs would have to be cleaned up, freight reloaded and broken cross arms repaired. At various times, tractors and sleighs might get stuck in the slush holes or muskeg and then they would have to be dug out. In hilly country there was always the danger of sleighs jack-knifing on a hill. During the first winter freight hauling season of 1938-1939, Carl Robinson, one of the swing bosses, experienced a jack-knifing accident going down a steep portage during the night. The tractor he was operating, as well as the sleigh loads of freight, jack-knifed and upset without his being able to jump clear of the wreckage. As a result, he was seriously injured.

Fortunately, a tractor swing returning from the mine came by shortly after the accident and took Carl back to Berens River in their caboose. It was rough enough for a healthy man to travel over a rough trail, never mind one that was badly injured. Upon arrival at Berens River, he was picked up by a Wings Limited plane and flown to a hospital in

Winnipeg. He recovered completely and was ready to take his usual captain's duties on the *Canvul* at Ear Falls, Ontario in the spring of 1939 when the navigation season opened. From the time this accident occurred, the steep portage was known as "Robinson Hill" and drivers used every precaution when crossing it.

Mel Parker related his own jack-knifing experience on Robinson Hill during the winter of 1944-1945. He said it happened near the end of the graveyard shift, when the sun was not yet clear of the horizon and the tractor lights did not illuminate the road very well. As Parker told me later:

"It was tough going that night, so we decided to 'double-up', the first and second tractor sleighs, making up a six sleigh load. The second tractor went ahead of my lead tractor with a 20-foot cable from the ring on the front of my tractor to his drawbar. We were travelling along in high gear across a portage when, all of a sudden, the lead tractor dropped out of sight. I knew right away what hill we were on but it was too late. I shut the throttle off and applied the brakes, but the six sleigh load behind was too much to hold, so I started to overtake the lead tractor. All that he could do was to try to keep ahead of me, but by that time my tractor went sideways so we were going downhill sideways, along with the six loaded sleighs on the drawbar of my tractor. Luckily, the bolt broke that held the cable from the lead tractor on the ring on the front of my tractor, swinging it around and crashing into the bush

Caboose fires left little to salvage.

and we came to a stop. Fortunately, no one was hurt."

Mel Parker was also in a serious caboose fire. It happened on one of his return trips to Berens River with a load of lead concentrates. The crews worked in eight hour shifts with breaks every four hours for coffee and snacks. Parker said that the on-shift brakeman came in the caboose at about 3 a.m. to make coffee and get lunch ready for the 4 a.m. break. He thought that the light in the Coleman gas lantern was out but there must have been some flame because when he took the filler cap off to fill the lantern, it caught fire.

The brakeman had a two gallon can of naphtha gas on the floor with the cap off ready for filling the lantern. When the lantern caught fire, instead of throwing it outside, he dropped it on the floor and tried to stamp out the fire with his feet. In doing this, he knocked over the two gallon can of naphtha gas and it only took a second for the whole floor to be covered with burning gas.

The brakeman then shouted and dashed out the door. Luckily the cook heard him, saw the fire, yelled and he also dashed out the door. Parker says he doesn't know who was next, but when he woke up, the first thing he remembered was seeing a can with a blue flame coming off the floor right from the bunks to the door and, as the flames went up the walls, they turned into a red flame curling to the back and funneling out the door and window. He looked at the small window at the front of the caboose but decided against going through it as he would go straight down between the moving sleighs. Instead, he made a run for the door.

"This all took place in a matter of seconds," he explained. "My trousers, etc., were right at the head of my bunk but I never thought of them as the only thing a person thinks about in such a situation is getting out alive, which I did. I never really woke up until I hit the road with bare feet and singed hair."

Parker continued: "By now, the lead tractor was nearly a mile ahead of us and when the driver

This tractor went through the ice at Pine Lake on the first trail breaking trip to Berens River in 1939.

stopped at 4 a.m. for coffee break, he swung his tractor around to come back to the caboose. All he could see was flames shooting up from the caboose, which he said looked like they were twenty feet high. In the dark he could only see our tractor's light moving and felt that we had all perished in the fire, which gave him an awful feeling. However, when he came back as far as the perishable caboose, he was very glad to see us all alive."

Another close call involved the time that Harry N. Everett was the tractor swing boss in charge of the winter freighting operations on Lake Winnipeg. One morning in January 1945, he went through the ice near Birch Point. Luckily he bobbed up in the open hole and was pulled out by his crew. Everett had an interesting insight into the experience. He said that he had always heard that it was easy to jump free when a tractor broke through the ice. But he said that from personal experience he could assure us that you don't have any time to react. The tractor goes down so fast that it is impossible not to go down with it. Your only chance is to come up in the hole, and then you might be pulled to safety by the crew. If you bob up under the ice, you die.

Chapter Four

A Winter Adventure

Opposite page: Tractor swing on the Black Birch River.

In the winter of 1935, I had the opportunity to accompany a tractor swing on a long trip from West Red Lake to the Woman Lake area. I use the word "opportunity" with some irony, as this was a trip in which everything that could possibly go wrong, did.

The preceding fall, the owners of the West Red Lake Gold Mine, located 20 miles west of Red Lake, decided to abandon the mine because of declining ore values. The Corless Patricia Gold Mines of Toronto purchased the West Red Lake Mine operation and contracted the Patricia Transportation Company to dismantle the entire plant and deliver it approximately 100 miles to their site at Woman Lake. Our company's quotation for this project was "cost plus 10 per cent" for dismantling the plant and a flat rate per ton for hauling the mine machinery, equipment and miscellaneous supplies.

Since I already had billing experience computing the weights of steel, pipe, angle bars, plates, rods, drill steel and the like, I was selected to proceed to West Red Lake Gold Mine property with the dismantling crew to carry out the computation of the freight bill. The mine supplied the company with copies of railway freight bills covering the heavy items, such as the steam boiler, hoists, et cetera from when this equipment was originally transported to the site. This only left the balance of plant equipment and supplies for which the weights had to be computed. It was estimated that the total tonnage to be transported was between 40 and 60 tons.

Instead of flying me to Red Lake, however, Charlie Wilson, managing director of the company, felt that I should have firsthand experience in winter freighting. He suggested that I should accompany the fully loaded tractor swing to the West Red Lake Mine property. Then, after the dismantling was completed and equipment loaded, he directed me to proceed with the tractor swing to Corless Patricia property, deliver the cargo and return to Hudson with the sleigh swing. This trip turned out to be an experience I'll never forget.

Company buildings at Hudson 1939.

The usual plan was to send a swing of two tractors with full sleigh loads of freight and a caboose from Hudson to Red Lake. However, on this trip, an extra caboose was taken along to be used by the dismantling crew at the mine. After the freight was delivered to Red Lake, one tractor with empty sleighs and caboose was delivered to Hudson. The other tractor was to proceed to the mine property with empty sleighs.

In the latter part of December 1935, the sleighs were loaded with approximately 50 tons of freight and readied for the trip to Red Lake. In order to avoid the Narrows, where the ice was unsafe due to the fast flowing water underneath, many portages had to be crossed. This delayed the trip somewhat, but we knew that if no slush conditions were encountered on the lakes, a round trip to Red Lake from Hudson could be made in a week. If slush conditions were encountered, it could take as long as three weeks to make the round trip. In other words, tractor freighting in the north was an adventure.

Generally, slush conditions are caused by an early freeze and a late snowfall. A heavy snowfall forces the ice down with its weight, and water comes up through cracks in the ice, turning the snow on top into slush. This slush freezes on top and when additional snow falls on top of the slush, the condition repeats itself. It is not unusual to have three feet of mixed ice and slush over three feet of solid ice. Another reason for slush conditions in this region was the fact that Lac Seul was a reservoir with water being released periodically at the Ear Falls Dam. This often caused the water level to drop and created a rough, broken ridge of ice at the approaches to the portages.

Early on the morning of New Year's Eve, 1935, the fully loaded tractor sleighs left Hudson. I was aboard, armed with a typewriter, calculator, freight bills, legal pads, a Ryerson's Steel catalogue showing weights, as well as a J.H. Ashdown Hardware catalogue showing shipping weights on different items. I was assigned the top bunk in the caboose where I was to try to sleep during the night on the 135 mile trip to Red Lake. By occupying the top bunk, I was also out of the crew's way when they changed shifts.

Ernie Wright and Arvid Anderson in 1941.

Up at the head of the train were two 40 HP Cletrac diesel crawler tractors. The crew consisted of drivers Ernie Wright, Clarence (Ole)

Erickson, Norman Howard and Alex Anderson; brakemen Arvid Anderson and Harold Johnson and cook Paddy Cochrane. It was a nice clear day with the temperature about 20 below zero. Fortunately there was no slush on Lac Seul, and with the clear weather, it looked like fine conditions for a good trip.

I assumed that all there was to winter tractor freighting was to get on a tractor, sit down in the driver's seat, put it in gear and move out with a string of sleighs in tow. However, in real life this is not the case, as the driver has to watch out for road conditions and know how to handle his sleigh loads. The hauling operation gets very complicated when you reach the first portage. Crossing steep portages with 50 tons of freight is an art.

When the tractor swing reaches the foot of the portage, it comes to a full stop. The brakeman unhooks the last two sleighs of the lead tractor's three sleigh loads. He then unhooks the second tractor from its sleigh loads and that tractor proceeds in front of the lead tractor where the brakeman fastens a 20 foot cable to its drawbar and to the front steel ring of the lead tractor. Then both tractors go up the steep portage towing the lead sleigh. This is known as "doubling-up".

When the two tractors reach the top of the portage, the brakeman unhooks the cable from the lead tractor. The second tractor then goes behind the sleigh and the brakeman hooks the cable to the rear bunk of the sleigh and to the front ring of the second tractor. Before going down the portage, the brakeman also puts a rough lock (logging chain) on the rear runner in front of the sleigh's bunk. This acts like a tire chain and slows down the sleigh as it slides down hill.

After this manoeuvre is completed, the lead tractor, in low gear, guides the loaded sleigh down grade while the tractor at the rear also proceeds in low gear, holding back the sleigh to prevent it from going too fast. If this procedure is not followed, the lead tractor can't control the sleigh load and it might push the tractor over and upset it along with its loaded sleigh. This is known as jack-knifing, and it can be very dangerous if the driver does not have a chance to jump clear.

When the bottom of the portage is reached, the sleigh is left behind on the ice and the two tractors go back and bring the rest of the sleighs across in the same manner. After all the sleighs and cabooses are across the portage, they are assembled again and hooked up to their respective tractors and the caravan continues.

The first day out, I became seasick, so to speak, from the caboose continually pitching forward and backward. The caboose pitched so much because the sleighs were dragging in the snow and the caboose at the rear had to go over an endless series of humps. So I went out and stood on the tractor's drawbar and hung onto the tractor's seat. On the first day of this trip, I mostly rode the drawbar of a tractor and had very little to eat because my stomach was so queasy. The next day, New Year's Day, was uneventful and as I was finally getting used to the constant pitching, I did not have to spend as much time standing outside on the drawbar.

The tractor swing reached the junction of the winter road leading to the hydro dam at Ear Falls when it was time to change shifts. We had to advise the Hudson office that we had cleared Goldpines and were heading to Red Lake, so I volunteered to go to the Ontario Hydro office and have them send a message by their radio operator to Sioux Lookout. Hydro relayed it to the Canadian National Telegraph office at Sioux Lookout and they in turn sent the message to their Hudson office which delivered it to the Patricia office. This seems like a roundabout way of doing things but in those days it was the only way you could relay a message.

Before I left the caboose for the two mile walk to the hydro office, Alex Anderson asked me what I wanted for supper, so I told him I wanted six fried eggs, sunny side up, six sausages and some canned peas, along with bread and coffee. He was surprised that I asked for so many eggs and with the usual charming manner of camp cooks everywhere in the north country, he said that if I didn't eat it all, he would shove it down my throat.

I was getting pretty hungry after having been seasick for two days. And it was a four-mile round trip walk in 20 degrees below zero temperatures, so I figured I could eat a horse. When I sat down to the table Alex had my meal ready and to his surprise, I ate the whole thing.

During that night, on crossing one of the portages, I was awakened to a lot of hollering outside and, before I knew it, the caboose flipped onto its right side. There was an enormous crash of dishes flying and the lights went out. We were all thrown from our bunks and the caboose filled with smoke. There was nothing I could do because I was trapped in the bunk at the top of the wall of the caboose. There was bedlam below. Arvid Anderson was trying to get up from the lower bunk and while he struggled, he caught hold of a couple of pairs of boots that were hanging on the side of the upper bunk. Feeling the dangling boots, Arvid shouted: "My God, somebody's legs are broken."

Harold Johnson, the brakeman on shift, pried open the caboose door and threw out the stove with its burning fire. Quick-witted Ernie Wright rushed back from the tractor and grabbed a couple of tractor batteries and threw them out. It was a good thing he did this, or we probably all would have fried to a crisp.

The caboose was righted in about 20 minutes and we cleaned up. It was a mess, with canned goods all over the place, the eggs thoroughly scrambled, pies turned into fruit cobblers, the floor soaking wet from the upset drum of water, as well as some oil upset from some cans. Everything was covered with soot from the knocked down stove pipes. It took about three hours to get things cleaned up and organized again for the tractor swing to continue on its journey to Red Lake.

On the evening of January 2, 1936, we arrived at Red Lake. Early the next morning, the freight was unloaded. The crew of one tractor sleigh and caboose returned to Hudson, while the dismantling crew for West Red Lake Gold Mine left for the mine property with one tractor, empty sleighs and caboose. Four men, myself included, were involved in this part of the operation. It was estimated that it would take from four to five weeks to carry out the dismantling and loading operations. There were no means of communications at the time from the West Red Lake property to the town of Red Lake, so a chance had to be taken that we would be through with our dismantling work at the end of five weeks.

One might wonder why it would take so long to complete the job. However, bear in mind that the work had to be carried out in subzero

Company Characters

Tim McGinnis

For some reason the profession of cooking seems to attract misfits and unusual characters. Tim McGinnis was a bachelor who was employed by the company as a cook at Root Portage. He also worked as a cook on the *Wapesi* during the summer, and as a caboose cook during the winter.

Tim would often let people know, in a rich Irish brogue, that he was an Irishman from Ireland. However, the closest that he had come to Ireland was Fredericton, New Brunswick, where he was born. Tim also placed a high value on his independence and vowed that when he got old, he would shoot himself rather than put up with the aggravation of all the aches and pains associated with the elderly.

When Tim reached the age of 75, he began to feel those creaky aches and pains. And true to his word, he went right ahead and shot himself.

"Five Thousand or Freeze"

When tractor trains are the only means of transporting food to hungry working men in the wilderness, it's necessary to calculate exactly how much food is needed for months at a time. We determined how much food would be needed in the remote camps by using a basic army formula, and adding a certain percentage to cover the increased consumption of calories in cold weather.

In the far north, the formula had to be increased simply because it was so cold. For example, Sir Peter Scott, Grandson of Captain Robert Scott, made the same trek to the South Pole in 1985 as his father had in 1911, and he consumed an average of 5,100 calories per day (which was less than the 8,000 calories daily the experts suggested).

The basic army ration for a soldier is a pound of meat, flour and potatoes per man-day. In reviewing the actual consumption of rations on far northern projects, we determined that we were feeding our men 1.52 pounds of meat products per day, one pound of flour, and about one pound of potatoes — higher than the army average, but necessary to keep the men functioning in cold weather.

weather with snowstorms in between. One can't work outdoors for very long periods in the cold weather. As well, daylight hours were reduced during the winter. Another reason for the slow progress was the problem of trying to find different items buried in the snow, which necessitated constant digging to locate items such as drill steel, steel pipe, rods and steel plates. In a little over four weeks the work was all completed, the buildings stripped of equipment and supplies, with the exception of the cookery, which was in use until the end.

While waiting for the Hudson tractor crew to arrive to help move the items, a mild spell occurred which opened up the narrows about a half mile below the property over which we first came in. We felt that we should warn the crew that the narrows were no longer safe. Therefore, we cut a row of evergreen trees and laid them across the intended route as a signal. We also posted several signs warning them of the danger. Fortunately, Ole Erickson was on the lead tractor, so he slowed down and his headlights shone on one of the notices. As a result, he came onto the mine property over the land portage. They arrived at about 4:30 a.m. and roused us, by announcing: "You sleeping beauties are enjoying your sleep while we have to work all night," along with various other unprintable comments. Needless to say, we did not get any more sleep that night.

Paddy Cochrane prepared breakfast for everybody and at daybreak we began removing the rest of the West Red Lake Mine plant. One crew

The crew which delivered this boiler to the Corless Mine in 1936 are, left to right: Norman Howard, Carl Johnson, Jack Wish, Baldy Dewey, Ernie Wright, Eddie Thompson and Ole Erickson.

was assigned to move the loaded sleighs individually over the land portage to the lake where they would be assembled for the tractor swing. Another crew loaded the steam boiler and other heavy equipment onto sleighs to be ready for the crew hauling them to the lake. A third crew stripped the cookery and loaded its equipment and supplies. After all the sleighs were hauled to the lake and assembled, the tractor swing with its cargo of 64 tons embarked on its way to its new home.

It was the middle of the night by the time we arrived at Red Lake, so at daybreak we went to the Red Lake Hotel for breakfast. Before going for breakfast, I contacted Earl Wilson, the company agent, to come and join us. I handed him a packet containing copies of my freight bills along with the time spent in dismantling the West Red Lake Mine plant and requested that he airmail this large envelope to Hudson as soon as the post office opened. I also requested that he send a message to the Hudson office via the Ontario Forestry radio system that the tractor swing with Corless Mine's freight was on its way to Goldpines.

We left Red Lake after a hearty breakfast and once again the journey continued. That night on Lake Pakwash a near disaster occurred. The tractor swing with all those heavy loaded sleighs was so heavy that the lake ice began slowly sinking. The sleighs weren't breaking through but the ice itself was actually being depressed by the weight of the tractor train, and there was a danger that water would flood onto the sinking ice and we would be unable to continue. There was a lot of hollering and a mad scramble as the crews woke up to help with the situation.

The crews uncoupled the sleighs hurriedly and pulled them in different directions. And when daybreak came, you could see sleighs scattered all over the lake. Twenty-foot cables were used to haul each sleigh load to the lakeshore road and there the tractor swing was reassembled. This was a slow process as great care had to be exercised to prevent any cargo from going through the ice. It took practically a whole day before the sleighs were all assembled and once again we were ready to embark for Goldpines.

We arrived at Goldpines in the early evening. The next morning, Alex Anderson and another driver and two brakemen left Goldpines by tractor with some empty sleighs and a caboose for Hudson, while our tractor swing — consisting of two tractors and sleighloads of freight — proceeded to the Corless mine property. The trip wasn't too bad over the 17 mile Bluffy Portage, but when the swing got to Bluffy Lake, we once again encountered a big problem — slush.

By late afternoon our sleighs were hopelessly stuck in deep slush. We were still about five miles from the mine and since I did not care to hang around to see them dig the sleighs out, I told Ernie Wright that I would walk over to the mine. As it turned out, I arrived at the mine at a most opportune time because the mine crew were just being

Opposite page: Sleigh stuck in slush on Lac Seul in 1935.

served supper. The mine foreman kidded me about being the "point man" for the tractor swing and wondered how far it was behind me with their loads. I estimated that the tractor swing would probably not show up at the mine until noon the next day as they had to dig each sleigh out and reassemble them in a dry area.

That's almost exactly the time they arrived. Clarence (Ole) Erickson asked the mine manager where he wanted the different equipment unloaded so as to be near the mine shaft. The mine foreman said, "Just dump the loads anywhere that's convenient."

Ole replied, "Surely you must have some idea where you are going to put up the steam boiler building? We'll unload the boiler there and it will just be a matter of skidding it a few feet onto a concrete base." Ole added that he could very easily spot the sleighs in each proper location, putting the water tank near the tower, the hoisting equipment near the shaft, and so on. But the mine manager again said, "Just dump the loads wherever you like."

This was rather strange attitude on the mine foreman's part but, it was pleasant news for the tractor crews. So they dumped the loads right where they sat. All they had to do was hook the cable on one side of the sleigh rack, put it across the other side, hook the cable to the tractor's drawbar and pull, which dumped the entire sleigh load. It did not take long to dump the loads and once again we were on our way back to Goldpines and Hudson.

It was towards the end of March when a warm spell caused the snow to melt and fill in the tractor trails on Lac Seul with water. Although the trail was filled with water, the tractor drivers stayed on it, as they knew the ice was still strong enough below to carry the weight of the machines. Ole Erickson came to an area that he thought looked bad and stopped the tractor.

Ernie Wright poked his head out of the caboose and wondered what the heck was holding him up. Ole replied that it looked bad ahead and that he felt he should go around it. Ernie told him that it looked fine to him and that he should go on. Ole remarked that he didn't want to be responsible for the tractor going through the ice with the empty sleighs, but if that was Wright's decision, he'd abide by it. He put the tractor in gear and stood on top of the tractor seat to be ready to jump if the machine broke through. Fortunately, nothing serious happened in crossing the area. However, when the caboose was going over it the water bubbled up through an airhole to its floor and Ernie conceded, "By God you're right, it's bad."

We finally arrived at Hudson about the third week of March, after almost three months on the trail. The next day, Charlie Wilson told us that we came very close to doing all that work for nothing. He had got wind of a rumor that Corless Patricia was in a financial bind. Upon receipt of my freight bills via airmail from Red Lake, Charlie took the train to Toronto and went straight to the Corless office with the bill. He informed them that the cargo would be delivered to their mine site in the next few days, and since the company crews had spent almost three months on the project, he wanted to pay them off right away. They gave Wilson the cheque for full payment and he made a beeline to their bank to have it certified. He then went to the Imperial Bank and deposited it in the Patricia Transportation Company's account. As it turned out, Corless Patricia Gold Mine folded before he left Toronto.

After hearing this, Ernie and I understood why the mine foreman had told Ole Erickson just to dump the loads anywhere. No doubt he and the rest of the mine personnel knew that the mine was going to be abandoned. To this day, I don't know if all that equipment was salvaged or whether it just sank into the muskeg and rusted away.

Chapter Five

The Early Days of Aviation

It would be difficult to overstate the importance of aircraft in helping develop the mining industry in the Patricia District.

Prior to World War II, aviators had the distinction of flying more freight tonnage to the northern mines and merchants of the Hudson area than was flown from either Heathrow Airport in England or the airports of New York. A magazine journalist called Hudson the "biggest little town in North America".

In the spring of 1924, the Ontario government set up its own flying service called the Ontario Provincial Air Service, which later became part of the Forestry Department. Shortly after, the Howey brothers and Jack Hammell decided that it was best to begin exploration work on their gold discovery claims in Red Lake. It was necessary to bring in men and equipment during the late fall of 1925 so that work could proceed during the winter months. There was no commercial air service that could do the job, so Jack Hammell persuaded the provincial air service to carry out the transportation. They used five Curtiss HS-2L flying boats to move the men and materials before freeze-up. J.R. Ross was the senior pilot in charge of the operation. He was assisted by R.C. Guest, H.A. (Doc) Oaks, T.G.M. Stephens and Rameo Vachon. When James A. Richardson's Western Canada Airways was formed, Ross and Vachon went over to fly their aircraft.

In 1926, a pilot named Jack Elliott of Hamilton, Ontario, heard about the Red Lake gold rush and decided to go into the transportation business. He loaded two Curtiss JM-4 aircraft into a railway box car and shipped them to Sioux Lookout where he assembled them and flew them to Hudson. From there he began the first air service to the Red Lake area on March 3.

The rate for 100 air miles was $1 per pound, adopted for both passengers and freight. The only passengers who could afford the service were mining engineers, promoters, and reporters and

Opposite page: Curtiss HS-2L flying boat at Kapikik Lake in 1929.

photographers. In spite of the high passenger rate of $200 for a 200 pound man, there was a queue of men at Hudson waiting to be flown into the Red Lake gold fields, so Elliott's two planes were kept busy every day. Passengers weighed in at the time they purchased their ticket and again before they boarded the airplane as it was not uncommon for the passengers to stuff their overcoat pockets with some essentials. They had to pay for the extra weight if they weighed more than when they bought the ticket.

Before Jack Elliott started to fly his aircraft, he had Elliott Brothers of Sioux Lookout make him a pair of plane skis from sketches that he furnished. These were the first skis to be made by the firm, but they later became well known for manufacturing them. Jack Elliott also devised a canvas nose hangar for use in servicing the aircraft during the winter months. This is believed to be the first time that such a hangar was used in Canada.

Aviators encountered problems with equipment in bush flying operations that were not normally faced by the air transport industry operating in the United States and Europe. One of the main problems was the type of skis provided by the aircraft manufacturers, so they adopted the solid type of pedestal with skis made by the Elliott Brothers, and these skis became well known throughout Canada.

Flying companies came and went in the early days of aviation. Many small bush flying operations were hampered by skimpy financing and overworked fliers. An example of this was the Patricia Airways and Exploration Limited of Red Lake. Doc Oaks, S.A. Tomlinson and Dale Atkinson purchased a Curtiss Lark aircraft and began air service on April 14, 1926 from Goldpines. The Lark was constantly in service, without much chance for proper maintenance. This situation not only hindered operations in the area but also jeopardized the lives of the crew. One day in front of Goldpines, the plane turned a complete somersault upon land-

ing. And since the cost of the repairs was more than the company could afford, that accident spelled the end of the firm.

Another flying company that was born of the mining industry was General Airways Limited. This company was formed in June, 1927 by well-known Canadian World War I ace Roy Brown, who many believe shot down Von Richthofen, Germany's leading fighter pilot. It originally operated out of Amos, Quebec, but with all the aircraft freighting activity in the District of Patricia, General Airways established a base at Hudson in 1935, with two aircraft to carry out the passenger and freighting service in the area. Tim McCoy was general

A Norrduyn Norseman IV, operated by Patricia rival Starratt Airways and Transportation, at Red Lake in 1936.

Opposite page: A Norseman BDC at Hudson in 1937, picking up freight for the Madsen Mine.

manager. Throughout the company's three years of operations, they carried a considerable number of passengers and handled large tonnages of freight.

McCoy was an excellent pilot but he had a short fuse. This was demonstrated one time in Red Lake when he was flying freight to McKenzie Island. A local citizen at the dock was loudmouthing about how great the bush pilots of the other companies were. McCoy, simmering inside, asked this gentlemen if he cared to come along for the ride to McKenzie Island. The man agreed.

On the return flight, McCoy decided to do a little "flying": he opened the throttle, looped the loop, flew under a hydro transmission line and generally put the Gull Wing Stinson through its paces at top speed. By the time they landed at the Red Lake dock, the passenger was white as a ghost, but he never again made unfavourable remarks about McCoy's ability to fly an aircraft.

One winter morning in 1938, McCoy took off from the Hudson base to pick up a passenger at Sioux Lookout. Before he was fully airborne, a strut on his Gull Wing Stinson broke, and because he did not have sufficient altitude, McCoy crashed in the bush and was killed. As a result of this crash, the Canadian government grounded all Gull Wing Stinson aircraft across Canada until the structural problem was modified. With no aircraft to replace the Gull Wing Stinson, the company's head office in Amos, Quebec ceased operations in the District of Patricia.

The Hennessy Airlines Company established a base at Hudson in 1937 and hauled large volumes of freight to the northern mines. They had two aircraft in operation but evidently they must have experienced cash flow problems as they only stayed in business one season and then left the area.

In March of 1928, Jack Hammell formed a successful company called Northern Aerial Minerals Exploration, with James A. Richardson providing 25 per cent of the capital. The company was also known as NAME. On March 20, Doc Oaks

ODE TO A FAIRCHILD

The Eighty-two was either loved, tolerated, or just plain detested by pilots. Charlie Robinson, one of Starratt Airway's pilots, had a hobby of writing poetry while in his Fairchild 82. Robinson, as it so happens, was very fond of the Fairchild 82, and during one of his flights out of Hudson to Red Lake in a CF-AXG in August 1939, Robinson composed this ode to his favourite plane:

A Pilot's Soliloquy

By Olden Bawld

I'm an ordinary pilot of a Fairchild Eighty-two
But there's no limitation to the things that I can do!
I rise and shine 'most every day right at the crack of dawn,
Before the other pilots wake I'm in the air and gone.

You never hear me grumble, I'm always bright and gay
Regardless of the weather, or the loads, or time of day.
There's fuel oil, grub and dynamite, and baled hay by the ton
I fly it all with joyous heart, and work from sun to sun.

I'm really in my glory loading double decker beds
And diamond drill equipment too, with greasy swivel heads.
I like to see the grease ooze out upon the cabin floor
And mix with lime, or flour, split from loads I've had before!

I never bounce my landings and I never bump the dock
I've flown on floats for years and years and never scraped a rock!
At forty-five degrees below and when I'm in a rush
There's nothing I like better than a pot-hole full of slush.

I never carry overloads, for that would be unjust
And anyone that says I do has got a lot of crust!
Neither do I fly at night; I never break Air Regs,
If that I'm ever guilty of I'll drink the bitter dregs.

There's never any argument twixt engineers and me,
I never cause them any grief, we're chummy as can be.
They always smile politely if I say the engine's rough
They never think I'm full of hops, or say 'It's good enough!'

I enter in the aircraft to log each trip in detail true
And never make the blunders that the other pilots do!
The Civil Aviation chaps all think that I'm OK
They write me nice long letters just to help me on my way.

I'm paid a lot of money just for flying through the air;
In five or six or seven years I'll be a millionaire!
The boss and I are buddies, he treats me like a son
And never fails to thank me when each flying day is done.

But time draws on, and when I'm old, I think 'tween me and you
I'll miss the joy and comfort of my Fairchild Eighty-two.
And though I'm just a pilot, I think it's really swell
And everything I've said is true
Oh yes it is...LIKE HELL!

left Western Canada Airways (another Richardson company) and joined Hammell as director of aerial operations and assistant manager. Their aim was to establish a company that could carry out extensive mineral exploration in northern Canada by flying prospectors to areas where gold might be found.

Their best gold discovery occurred on the Crow River in the Pickle Lake area of northern Ontario, and, of course, in the District of Patricia, where Pickle Crow Gold Mines was established and operated for 31 years. This find was about seven miles from Pickle Lake and led to the establishment of Central Patricia Gold Mines, which was about one mile from Pickle Lake. It operated for 18 years before the ore bodies were depleted.

Starratt Airways and Transportation Company Limited began in August 1932 when R.W. Starratt purchased a CF-AGX Gypsy Moth to keep track of his winter tractor swings. By 1937, the company had a total of ten aircraft operating in the District of Patricia. During the winter of 1940, a Beechcraft CF-BGY piloted by Starratt's son Bud crashed on Lake Pakwash on his return flight from Red Lake to Hudson, killing Bud and a passenger. The cause of the crash was found to be a break in the heating line, which caused carbon monoxide to leak into the cabin. Bud inhaled the odorless, tasteless gas and was rendered unconscious, at which point the plane crashed and he was killed.

Charlie Robinson logged many hours flying a Gypsy Moth out of Hudson before he was assigned to his favourite Fairchild Eighty-Two. On one of his flights, he had to avoid a thunderstorm so he landed on Bending Lake on a sandy point by the entrance to the Turtle River. He liked the spot so much that he figured it would be an ideal place for a dream home when he retired.

After R.W. Starratt sold his company to Canadian Pacific Railway, Canadian Pacific Airlines transferred Robinson to the Yukon to fly materials for the Dew Line. In 1943, he was the first pilot to land a Boeing 247 on wheels in Yellowknife, Northwest Territories. After leaving Canadian Pacific Airlines, he went on to fly various aircraft for Trans Canada Airlines until his retirement in 1962 just before the advent of jet aircraft.

Upon his retirement, Robinson went back to the dream spot on Bending Lake, built a home and lived there with his wife Rea. They were miles away from the closest road and Charlie flew his Cessna 182 to Ignace to pick up the mail and groceries.

Ignace held its first annual bush plane air show on September 6, 1989, where aircraft of the bush flying era from all over Canada were in attendance. The special guest of honour was Charlie Robinson and his wife. Robinson delivered a speech about bush flying and he delighted the crowd with some of his flying poems.

A New Company

Western Canada Airways was an exceptional company and, with time, it became a major force in the Canadian aviation industry.

In the fall of 1926, H.A. (Doc) Oaks approached James A. Richardson, a well-known Winnipeg grain baron and financier, about establishing an aviation base at Goldpines, Ontario. It did not require much of Doc Oaks' persuasive abilities to interest Richardson, as he was vitally interested in the mining industry. Richardson thought that the introduction of aircraft would revolutionize development of mineral resources as prospectors could be flown to the area of their prospecting sites within a few hours as opposed to at least a week to reach their destination by canoe.

The new company was granted a charter on December 10, 1926, with wide powers of operations in the field of aviation. The head office was to be located in Winnipeg and the original capital stock, worth $200,000, was held by James A. Richardson. Officers of the newly-formed company were James A. Richardson, president; Mrs. James A. Richardson, vice-president; John Hunter,

Doc Oaks and his City of Winnipeg *Fokker Universal at Hudson in 1926.*

secretary; H.A. Oaks, manager, and J.A. MacDougall, treasurer.

Doc Oaks, accompanied by engineer Al Cheesman, travelled to New Jersey and purchased a Fokker Universal monoplane powered by a 200-hp. Wright J-4 Whirlwind engine. They dubbed the plane the *City of Winnipeg* and — on a freezing December day in 1926 — flew it from New Jersey to Hudson, Ontario. The aircraft was capable of operating on wheels, skis or floats. It was the first of a number of types of aircraft powered by the new air-cooled radial engines. These new aircraft changed the nature of air transportation across North America, particularly in the Canadian north.

Oaks immediately contacted the Elliott Brothers of Sioux Lookout and arranged to have them make new skis to replace those supplied by Fokker. Elliott Brothers' skis became standard equipment on Western Canada Airways aircraft. K.M. Molson in his book *Pioneering in Canadian Air Transport* related the following story:

"In the first week of 1927 it was decided to open an outpost at Goldpines on Lac Seul on the way to Red Lake in order to serve as a distribution point for freight to the outlying areas where the mining men were working. A.H. Farrington, who had learned to fly in World War I, was hired as the agent. He set up a tent and equipped it with a wooden floor and an oil stove for heat. From there he dispatched the first fur bales to travel by air on January 13, 1927.

On February 2, after leaving in the morning with 800 lbs. of freight for Narrow and Clearwater Lakes, Oaks disappeared. Two days later J.A. MacDougall, who had come down from Winnipeg to set up the bookkeeping system, reported to Winnipeg that Oaks was missing and that Cheesman had started up Lac Seul with a dog team and repair materials. The weather had been bad ever since Oaks left and it was hoped that he had merely landed to sit it out. Cheesman found the aircraft at Woman Lake, 122 miles from Hudson, grounded by a damaged undercarriage. Oaks, in the meantime, had been flown out to Hudson in a Patricia Airways' Lark, piloted by Dale Atkinson. He then wired New York for spare parts and boarded the train for Winnipeg."

John Andrew MacDougall, treasurer of Western Canada Airways, at Hudson in 1927.

James A. Richardson

Often described as "the father of Canadian aviation", James A. Richardson not only founded Western Canada Airways, but he also owned a myriad of other Canadian companies, including the Patricia Transportation Company. Richardson used his considerable foresight, experience and financial power to turn both companies from shoestring operations into solid companies with international reputations.

James A. Richardson was a daring entrepreneur who sought out business opportunities on the ragged edge of the Canadian frontier. The grandson of a grain merchant from Kingston, Ontario, James A. Richardson was still a child when his father re-established the family firm in Winnipeg in 1896. Winnipeg was the centre of the Canadian grain trade and the Richardson empire grew rapidly. After his schooling was finished, James A. Richardson could have easily managed the grain brokerage business and rested on the family laurels. But JAR (as he was widely known) enjoyed the risk and challenge of northern development projects. Under Richardson's stewardship, Western Canada Airways became the country's first coast-to-coast airline, and the Patricia Transportation Company grew into the premier wilderness freight-hauling firm.

James A. Richardson was a large, good-natured gentleman who enjoyed the company of working men. He was interested in the day-to-day problems faced by those who toiled in the wilderness, particularly the daredevil tractor drivers and pilots. Upon Richardson's death in 1939, the Canadian north lost one of its greatest pioneers.

James A. Richardson, second from right, at Winnipeg in 1938.

Freighting activities through the summer of 1928 were very good, especially beginning in July when a big boom developed at Favourable Lake. In this area approximately 250 miles north of Hudson mine operators were dependent upon aircraft for all equipment and supplies. C.H. (Punch) Dickins was kept busy on this operation with occasional help from Harold Farrington.

Since its first flights, the company carried airmail as a matter of convenience into the Red Lake area. Oaks was anxious to put it on a more businesslike basis and requested Richardson contact the postal authorities the next time he visited Ottawa. On March 4, 1927, the Post Office gave approval to Western Canada Airways to issue their own airmail stamps. Such stamps did not resemble any regular postage stamps or carry numerals indicating their value, but were authorized by the postal department. The company ordered 20,000 stamps from a printer in Winnipeg. On receipt of these stamps on April 9th, the Post Office determined that they were to be sold for 10 cents each.

In January 1928, Western Canada Airways was awarded the mail contract from Hudson to Red Lake via Goldpines, with one trip per week scheduled at up to 400 pounds. The inaugural trip was January 25th, made by Dale Atkinson with a load of 588 pounds of mail. The company was also awarded airmail contracts to carry mail into Narrow Lake, Pickle Lake and other mining camps. On March 14, 1928, Western Canada Airways was granted the contract for carrying mail from Hudson to Favourable Lake, a distance of approximately 250 miles. The flying goose insignia first appeared on Western Canada Airways aircraft in June of 1929. By the end of 1930, the company controlled almost all the air transport business in Canada.

In late March of 1937, Western Canada Airways was advised by the Post Office that on April 1st, the contract for carrying of mail into Pickle Crow was to be turned over to Starratt Airways. There were no complaints as to the service and the

A 2,100 pound ball mill in a Junkers airplane, en route to Central Patricia Gold Mines in 1936.

rate was 10 cents per pound. It was understood that Starratt was given the contract at 9 1/2 cents a pound, provided the mail ran 100 pounds daily. This was only a difference of 50 cents per day.

Apparently no consideration was given to Western Canada Airways for pioneering the delivery of mail to the area for free at the start of aviation in 1926 and being competitive in other years. It was later determined the contract was given to Starratt Airways and Transportation Company Ltd. for political reasons.

The Depression shook up the whole aviation industry, and most small airline companies in Canada went under. The shake-up resulted in several strong companies surviving. K.M. Molson explained in *Pioneering in Canadian Air Transport:* "Canadian Airways Limited came into being on November 25, 1930 when it took over the assets of the Aviation Corporation of Canada and Western Canada Airways Limited. Richardson signed the Bill of Sale for Western Canada to the new company on November 19."

James A. Richardson, whose initiative and energy was largely responsible for the development of aviation in the north country, was the majority stockholder and president of the new company. The vice-presidents were Sir Henry Thorton, president of Canadian National Railway and, Sir Edward Beatty, president of Canadian Pacific Railway. Apart from Richardson, Thorton and Beatty acting as directors of the company, 13 others were also appointed — bank presidents and presidents of major Canadian companies. Richardson felt that it was important to have an impressive board of directors to convince American aviation interests that Canadian Airways was a strong company.

Things got off to a rocky start, nonetheless. As the Depression worsened, the profitability of Canadian Airways' operations declined considerably. At their May 1934 meeting, the directors ordered a 10 per cent reduction in pay to all personnel except the pilots, who would have the same pay, but would not receive their mileage allowance until they had flown 1,700 miles each month. Pilots Roy Brown, Jack Moar, Milt Ashton and Ted Stull were dissatisfied with this arrangement and they all resigned in July 1934 and formed Wings Limited, based at Lac du Bonnet, Manitoba.

The company was incorporated July 11, 1934 with Roy Brown as president and Milt Ashton as general manager. Their licensed routes in the District of Patricia were Sioux Lookout to Red Lake and Pickle Lake; Winnipeg to Red Lake and Kenora to Red Lake. The company made inroads into the business originating at Lac du Bonnett, but did not do so well in other areas, although they had 10 aircraft operating in the district in 1937.

The Flying Goose

The "flying goose" insignia found on all Western Canada Airways planes is an interesting tale in itself. The first goose appeared in 1929 and the insignia was passed on to Canadian Airways aircraft in October 1931. The stylized flying goose continued to fly in modified form for 40 years, and then was adopted by Canadian Pacific Airlines. So it was that the bird that once accompanied Doc Oaks and Wop May on their flights across the northern wilderness eventually found a home on the tail stabilizers of CP Air Jet Empress airliners flying across the world's oceans.

Canadian Airways emphasized safety and service. There are many stories that illustrate this, but a typical incident occurred in 1936, when the Central Patricia Gold Mines at Pickle Lake, Ontario had their ball mill break down. This mill had to be replaced as soon as possible or the milling operations would have to be suspended until the opening of navigation early in June. The only aircraft in Canada that could handle this large piece of equipment was a Canadian Airways Junkers 52-ARM. This airplane was operating out of Yellowknife, handling large tonnage for hydroelectric operations. Central Patricia Gold Mines wired Canadian Airways' head office in Winnipeg for help. Winnipeg immediately radioed Yellowknife for the Junkers. In the meantime, the eastern factory shipped the unit with bearings by Canadian National Railway express to Hudson.

Pilot Stuart McRorie took off from Yellowknife on April 8th, 1936, for the 1,500 mile flight to Hudson and arrived on April 9th. The ball mill unit weighed 2,100 pounds. The two bearings weighed 350 pounds. Each was loaded into the aircraft cabin by the Patricia crew and braced by 12" by 12" timbers. Upon completion of loading, the plane took off for Pickle Lake.

If Canadian Airways had not stepped forward to help, the Central Patricia mine would have been in a fix. The mine's milling operations would have been suspended for two months until the opening of navigation to Doghole Bay. As it turned out, the unit was installed at Central Patricia Mine and was operating by 4 p.m., just three days after the request was made to the Canadian Airways' office in Winnipeg.

Before the Junkers 52-ARM returned to Yellowknife, it stopped off at Goldpines and moved 175 tons of freight to Uchi Gold Mines at Uchi, a distance of approximately 50 miles. The freight consisted of lumber, angle iron, steel bars, iron pipe in 22 foot lengths, heavy steel plates, eight large electric switches weighing a total of 2,400 pounds, ten tons of calcium chloride, and more than 80 tons of cement and furniture for a good-sized house.

The crew of the **Wapesi** *listen to the radio broadcast of James A. Richardson's funeral in 1939.*

James A. Richardson suffered an untimely death, at the age of 53, from a heart attack at his Winnipeg home on June 26, 1939. It was not only a great loss to the company but to the aviation industry in general. He was frequently called the "father of Canadian aviation" and his death was a most severe loss to the company.

Mrs. James A. Richardson carried on with the company's operations and in 1940 Sir Edward Beatty became president, succeeding Richardson. G.W. Hutchins was appointed executive vice-president, a newly created position. The Canadian Pacific Railway Company approached Canadian Airways Limited with a view of acquiring the majority control of Canadian Airways and its subsidiaries held by the Richardson estate. A satisfactory agreement was reached regarding the value of the companies and the Canadian Pacific Railway (CPR) arranged to provide a cash settlement.

After the CPR directors agreed to these arrangements, Mrs. Richardson wrote a letter to G.A. Thompson, general manager of Canadian Airways, and to the managers of its subsidiaries — Arrow Airways, Dominion Skyways and Quebec Airways. She said that she had decided to sell the company to the Canadian Pacific Railway in the best interests of aviation. The sale became effective January 1, 1942 and Mrs. Richardson's letter was published for all the employees in the January 10, 1942 issue of *The Honker*, the company's monthly newsletter. By those events, the largest aircraft company in Canada passed into ownership of the Canadian Pacific Railway.

OF WINNIPEG

Chapter Six

The Bush Pilots

Bush pilots were a special breed of men who needed to have the same sort of courage and self-reliance as the first explorers of Canada.

In the early days of aviation in the Canadian north country, airplanes were not equipped with radios to communicate with their base or points of destination. The only available maps were badly outdated because they usually had been created by early explorers who only drew the rivers upon which they travelled. As a result, many nearby lakes were not recorded. Therefore, these pilots had to use their best judgment in locating their destinations and fly "by the seat of their pants".

At the start of northern aviation, most pilots were World War I veterans. There was also a small group of younger pilots seeking adventure in the air. Neither group was motivated by profit, as there wasn't much compensation in terms of pay for the many hazards they encountered.

One of the dangerous conditions the pilots encountered was known as "glass-out". Glass-out occurred when the water on the lake was so calm that it looked like a mirror. In the winter, a smooth coat of unbroken snow made it just as difficult to judge altitude when coming in for a landing. If a pilot misjudged his height, he might strike the ground in a nose-down manner and crash the airplane.

Any new northern pilot, no matter how experienced he was in southern aviation, had to be trained how to operate a plane in the north country. Part of this training involved learning how to manage an aircraft equipped with floats or with skis. Floats, for example, change the flying characteristics of an aircraft, making it heavier and more clumsy. Pilots had to relearn many of their basic lessons, and when flying an aircraft at 100 miles per hour, there was no opportunity to try again if the pilot made a serious mistake.

Bush pilots were not only required to handle the aircraft, but they also had to be capable of

Opposite page: Left to right: J.R. (Rod) Ross, Bernt Balchen, S.A. (Al) Cheesman and F.J. (Steve) Stevenson stand beside the City of Winnipeg *at Hudson in 1927.*

F.J. Stevenson was killed in the service of Western Canada Airlines at The Pas in 1928. Winnipeg's first major airfield was named in his honour.

looking after their passengers during any possible emergencies. Pilots were often called upon to provide air ambulance services, and they habitually made visual inspections of camps and trails to see that expeditions were proceeding safely. During the Red Lake gold rush of 1926, when hundreds of men were walking from Hudson to Red Lake to discover that pot of gold, the pilots posted notices instructing travellers to lay green spruce boughs on an open stretch of snow if anyone was in trouble. The sick and injured could then be picked up and flown to the Red Cross hospital railway car at Hudson.

There were many accidents in 30 years of aviation history in the Patricia District, and while pilots were killed, few passengers died. It is a testimony to the skill of those pilots that no fatalities were caused by the occasional mechanical failure. (In southern aviation, the great majority of accidents are the result of pilot error.) Pilot Tim McCoy of General Airways lost his life when the Norseman he was flying crashed just outside of Hudson in 1938 upon take-off. Bud Starratt of Starratt Airways and Transportation Company lost his life when the Beechcraft he was flying in 1940 experienced a broken exhaust line and carbon monoxide leaked into the cabin. Bud passed out and the plane crashed on Lake Pakwash, killing him and a passenger instantly. For the most part, the extraordinary skill of these pilots was earned in aerial combat conditions during the First World War. When James A. Richardson founded Western Canada Airways, he decided to seek out the most outstanding ex-military pilots in aviation. They included Captain Frederick J. Stevenson, who was awarded the Distinguished Flying Cross and the French Croix de Guerre; and, C.H. (Punch) Dickins and W.R. (Wop) May, who were presented with the Order of the British Empire (Civil) for being the outstanding Canadian bush pilots. They were also awarded the McKee trophy along with H.A. (Doc) Oaks. The other outstanding pilots that operated in the District of Patricia during those years were Bernt Balchen and J.R. (Rod) Ross, and last but not least, was the air engineer, S.A. (Al) Cheesman.

It is a matter of some interest that during Admiral Byrd's Arctic and Antarctic trips, (as well as Wilkins and Ellsworth's expeditions to the poles) crews were augmented by bush pilots whose expertise in flying in the Canadian north country made them invaluable to the success of these expeditions.

W.R. (Wop) May

ESTERN CANADA AIRWAYS LTD.

In 1928 pilots Doc Oaks and Bernt Balchen, along with air engineer Al Cheesman, were granted leaves of absence by James A. Richardson to participate in Byrd's Arctic expedition. Oaks recommended certain modifications be made to aircraft for flying in extreme cold weather conditions, especially in the design of the skis.

Apart from carrying passengers to established areas, bush pilots also explored new country. In charting a new area, prospectors would be flown into a promising place with their canoes, camping equipment and food supplies. If the prospectors were successful in locating an interesting ore body, a diamond drilling crew would follow to explore the extent of the ore body. They required small gasoline engines, many lengths of drill rods, gasoline for the engines and general supplies.

If the diamond drill results warranted, a mill would be established and, in some cases, all the necessary equipment would be flown in. Heavy equipment would be broken down in components small enough to go into the aircraft and be reassembled at the mill site. Many other items needed to be delivered, and these all went in by airplane–tons of dynamite, cyanide, cement, ball mill liners, steel balls, 45 gallon drums of fuel oil, and many other items, all of them heavy. For example, a complete set of ball mill liners weighed a total of 8,700 pounds. Mine hoists in sections weighed 1,400 pounds, and mine cars weighed 750 pounds each. In addition, such unlikely items as livestock were sometimes flown in by aircraft.

When Western Canada Airways first set up shop at Hudson in 1926, Bernt Balchen was put in charge of the flying operations. In the book *Pioneering in Canadian Air Transport*, Balchen described his first arrival at the corporate "headquarters":

"We follow the lantern down the railroad embankment along a narrow path and ahead of us I see the pink glow of a Yukon stove lighting the windows

Western Canada Airways' first office at Goldpines, Ontario in 1927.

of a snow-covered shack. It is little more than a lean-to but the lantern's rays reveal an impressive sign over the door–Western Canada Airways. This is the administration building, ticket office, freight station and passenger terminal for the whole flying gold rush.

"As we push open the door and stamp the snow from our feet a group of men around the pot bellied stove peer at us curiously through the murk of spruce wood smoke and stale tobacco. The warmth is welcome and my three mechanics make for the stove at once. One of them, a youngster from Jersey City, lifts his coatails and extends his rear end towards the heat, shaking himself gratefully. A short dark bearded member of the group shuffles toward me in beaded slippers and holds out his hand.

'I'm Rod Ross. Airways Superintendent, here. You're Balchen, I take it.' We shake hands and he introduces the others — 'This is Tommy Siers and Al Cheesman, our maintenance chief.' He jerks a thumb towards a lanky figure seated behind the stove, —'and that is Captain Stevenson, one of our pilots.'

"Stevenson is scrunched so far down in his chair that he is almost sitting on the back of his neck. His legs are propped on a high shelf, a pair of moose hide moccasins comfortably crossed and a curved pipe is hooked in his mouth, the bend of the stem following the line of his long angular jaw. He waves a hand languidly in greeting.

The young mechanic from Jersey is fidgeting beside the stove and his eyes move around the little room in embarrassment. He enquires in a low voice, 'Which door is the men's room?' Captain Stevenson unlocks his moccasin feet from the shelf, clambers to his full height and rips a page from the Eaton's Mail Order Catalogue hanging on the wall. He opens the outside door and points to the darkness and swirling snow. 'There's the whole wide world, Sonny Boy,' he drawls, 'and if you can't help yourself, you're no man for the north country.' "

On January 19, 1927, Doc Oaks received a wire from A.E. Godfrey of the Royal Canadian Air Force at Hudson requesting that he pick up men and material at mile post 350 on the Hudson Bay line near Cache Lake, Manitoba, to be transported to Churchill, Manitoba.

At nine o'clock the next morning the two aircraft took off for Cache Lake. Rod Ross was appointed to lead the expedition, Stevenson and Balchen were the two pilots and Al Cheesman their air engineer. Stevenson piloted Fokker Universal FU and was accompanied by Ross, while Balchen was at the controls of the Fokker GD with Cheesman. They made an overnight stop at Norway House located at the north part of Lake Winnipeg. Upon arriving at Cache Lake they found that the lake was small and not very suitable for operations, which would restrict the size of loads. No communications were available between Cache Lake and Churchill, which ruled out any means of checking on the weather or determining any problems the machines would have at destination. Furthermore, the original estimate of transporting a few men and one ton of materials grew to nine tons and fourteen men.

On March 22nd, both aircraft left on their first trip to Churchill. They landed at Rosabelle Lake about five miles from Churchill where they unloaded 1,200 pounds of freight, leaving Ross behind to survey landing conditions at the harbour. In effecting the delivery of the freight, they encountered all kinds of problems — extreme cold, aircraft breakdowns, and so on. However, they completed the delivery of men and freight on April 7th, and both aircraft returned to Hudson.

On their arrival at Hudson on April 22nd, (they still had skis for landing gear) they found that most of the lake was now open water. Spring breakup had arrived several weeks ahead of schedule. This was the final straw on a very difficult trip. But Oaks and Balchen found an area of the lake with sufficient ice along the shore to touch the planes down,

Harold Farrington and his Gypsy Moth.

then they kept their speed up and ran the planes onto shore.

Early in March 1929, Harold Farrington was departing from Red Lake for Sioux Lookout with a lone passenger on board, Gerald Rowan. Their takeoff was very rough because of large snow drifts on the lake and one of the plane's skis was completely torn off and left behind in the snow. There was no radio on the aircraft in those days, so a plane took off and flew alongside Farrington's plane to signal him that a ski was missing.

Farrington noticed, but decided to proceed to Sioux Lookout with his passenger. Red Lake radioed Sioux Lookout to have an ambulance ready on his arrival. However, he changed his course for Hudson as this was the area he knew so well. When he did not land at Sioux Lookout, the ground crew sent a wire to Hudson base and told them to be on the lookout and provide first aid as there was no doctor at Hudson and no road for an ambulance to be dispatched from Sioux Lookout. In those days the only way out of Hudson was by the passenger train that came through at midnight or by airplane, weather permitting. They also asked Hudson base to have a plane ready to take off as soon as Farrington's plane appeared, because it was very likely that the plane was going to crash or at least flip over when the pilot attempted a landing.

Farrington was a cool customer and upon arriving in Hudson, he circled several times. His passenger was still totally unaware of the situation and he asked Farrington why he was circling and not landing. Farrington pointed to the side minus the ski, and explained that they were liable to bump a little hard on landing.

He suggested that Rowan take an eiderdown sleeping bag from the rear of the plane and wrap it around his head. Rowan said that about this time he was ready to give up all his mining claims, as he did not think they would come out of this situation alive.

C.H. (Punch) Dickins.

Farrington made one last sweep over Hudson and then tried landing on one ski. Incredibly, he kept the plane skidding along on one ski until it was almost stopped, then it flopped down and bent the propeller. Otherwise, no one was hurt and nothing was damaged but the propeller.

The famous pilot Punch Dickins flew out of Hudson for several years, and he had many adventures while flying in the Patricia District and elsewhere in the north country. In 1931, Dickins was put in charge of the Mackenzie River District based at Fort McMurray in the Northwest Territories. In his April 9th report to the Canadian Airways office in Winnipeg he noted:

"The aircraft business of flying passengers and freight has been pretty good. However, I am having a little problem with the Accounts Receivable. There are no banks in the country and no cash to speak of as it is the policy of the two large trading companies to keep as little cash as possible in the country as it is adverse to them in dealing with the trappers and Indians when buying furs. When there is no cash, the sellers of fur are obliged to accept an order for credit at the company concerned and take it out in trade on which the company makes a profit.

"For example, on my last trip on April 3rd, I arrived back at Fort McMurray with $15.00 in cash; $240.00 in wolf bounties; $320.00 in beaver and destitute ration orders; 400 muskrats; two red foxes; two cross foxes; seven martens; five minks; one lynx and fifty extra muskrats to come and go on. It is difficult to see that there will be any change in the monetary system in the future as the main trade is still controlled by the two trading companies and I expect to have to carry on under the same conditions for some years."

On one occasion, Art Shade was flying loads of four 45-gallon drums of fuel oil per trip in one of the Canadian Airways Junkers to Argosy Gold Mines at Casummit Lake. On his return trip from Argosy, he spotted an Indian with his dogteam on Confederation Lake (formerly known as Woman Lake) on his way to Goldpines. He landed near the Indian with his dogteam, loaded them all in the Junkers and took off for the Goldpines base. He knew that there would be no one at the base to help him unload those 450 pound drums of fuel oil, so he offered to fly the Indian and his dogs to Goldpines in return for help with the drums.

Other Indians who were also heading with their dog teams to Goldpines at about the same time were quite surprised to see that their friend had beaten them to Goldpines by two days. They could not figure out how he reached the destination so quickly.

Jackie Bowman was a daring pilot. He was the son of Frank Bowman, a Hudson businessman who owned a general store and a fishing business. Jackie might be called reckless as he was always up to some crazy antics as a teenager. For example, he would climb to the top of a forestry lookout tower, hook his legs on a beam and with his head hanging down, he would taunt the other boys to try and perform a similar feat.

Flying was in Jackie's blood and he idolized the local bush pilots, so at age 16 he persuaded his dad to buy him a Piper Cub. Jackie canvassed the businesses on the waterfront for any errands that they might have so he could put in as many flying hours as possible. He also wanted to practice taking off and landing with skis in order that he could obtain his pilot's license.

One day Jackie came into the Patricia Transportation Company's office and asked Don McLennan if he had any errands that he could carry out. It so happened that earlier that day Ernie Wright had left with a tractor swing for Red Lake and he had forgotten some parts on the counter. McLennan told Jackie: "You want to do an errand, take these and deliver them to Ernie on his way to Red Lake."

Jackie immediately took off with the parts and after a couple of hours returned to the office and told McLennan that he had delivered the parts to Ernie when he caught up to him on a portage. McLennan asked: "How in the world did you deliver parts to Ernie on a portage as you could not land with skis?"

Jackie replied: "It was easy as I made a few low passes over the tractor swing and managed to talk to Ernie and dropped the parts off."

By 1936, Jackie finally put in the required flying hours, but to get his license, he had to go to Winnipeg's Stevenson's Airport to be checked out. Jackie's father offered to ship the Piper Cub by

The **Hexagon** *with Junkers transport at Hudson in 1936.*

freight to Winnipeg, but Jackie would have none of this. He was determined to fly it there himself. One problem was that Winnipeg was 252 miles away and the Piper Cub did not carry sufficient gas to make the trip. Everyone tried to discourage him, but Jackie said that he had the gasoline situation solved as he was going to take an extra five gallon can of gas and siphon it into the plane's tank in mid-air. He also said that he would follow the Canadian National Railway's track to Winnipeg.

Another little problem was that he would have to have wheels on his Piper Cub to make a landing on Stevenson Field. Lost Lake was not clear of snow so he could not take off on wheels in the snow. Still, Jackie wasn't worried. One nice clear morning he had the Piper Cub towed about a mile east of Hudson to where there was about a half mile stretch of straight road on the way to the Keewatin Lumber Company plant. From there, he took off on wheels for Winnipeg. His father was very concerned that he would not make it, but in spite of all the odds against him, Jackie arrived safely and received his pilot's license.

When the war broke out, Jackie joined the Royal Canadian Air Force and fought overseas. When he returned he became a captain for Trans Canada Airlines.

Young fellows like Jackie Bowman learned a lot from the many experienced ex-military pilots who had learned to fly airplanes during the First World War. One of these experienced pilots was Mike De Blicquy. In the summer of 1938, De Blicquy flew a considerable tonnage of fuel oil in 45 imperial gallon drums to Argosy Gold Mines in his Junkers. The Junkers was a good freight hauler and was capable of carrying six drums or a total of 2,800 pounds per trip. On one clear, calm day, the water on Lost Lake was very placid, with not even a ripple on it. This condition is what is known as glass-out. It is not only a hazardous condition for

Missing May

In the winter of 1940, after World War II had broken out, Bill May of Canadian Airways Limited was selected to fly Hollywood actress Mae West from Winnipeg to Churchill, Manitoba, to entertain the military personnel stationed there. It was May's responsibility to see that the party travelled safely to Churchill from Winnipeg.

The morning after her performance, Mae West boarded Bill's aircraft for the return flight to Winnipeg. The weather, according to May, was not very good so when they got about halfway to Winnipeg, Bill decided to land at the little mining town of Flin Flon until flying conditions improved.

His fellow pilots never gave up ribbing Bill about this flight, telling him that the weather wasn't so bad and he could have easily continued his flight to Winnipeg. They seemed to think that his "weathered in" story was just an excuse so that Bill could spend a couple of days in Flin Flon with the sexy Mae West.

Opposite Page: Crashes were not uncommon in the early days of aviation.

landing, but it also limits the payloads of planes because it creates suction on the floats and planes have difficulty taking off. So that day, the pilots had to wait until there was some breeze which would create some waves.

However, the glass-out did not stop Mike De Blicquy from moving his full payload of six drums of fuel oil. After the aircraft was loaded, he taxied from the dock and started bouncing the aircraft from one side to another to create waves, then — when the surface of the lake was disturbed — he turned around and took off in the rough water.

On his return trip, De Blicquy circled the lake at Hudson, threw a cushion out of the window so that when it landed on the water it created ripples. This gave him the visual depth perception for a safe landing. After landing and taxiing to the dock he came into the office and ribbed his fellow pilots that he was making money for the company while they were sitting around telling dirty jokes.

Of course, sometimes there were accidents. During the night of September 1st, 1939, John Drybrough, a mining engineer from Winnipeg, arrived at Hudson (along with Kenneth Carlyle) on the midnight passenger train. Drybrough was going to the Argosy Gold Mines at Casummit Lake to carry out some professional work, while Carlyle was on his way to Red Lake. The next morning a Canadian Airways Junkers 34, piloted by Roy St. John, arrived to pick the men up. Since Drybrough was going to Argosy, Charlie Wilson, managing director of Patricia Transportation Company, decided to go along and visit with the mine manager.

Although dense black thunderclouds were gathering ahead, the pilot felt that it was only a local disturbance in view of the favourable radio advice from Red Lake indicating clear weather. He decided to take off just before 7 p.m. Central Standard Time and fly above the cloud layer as it was only about a 30 minute flight. At this juncture, it might be mentioned that 7 p.m. might be consid-

ered fairly close to darkness, but in this part of the north country, at this time of the year, it is still daylight at 10 p.m..

After flying long enough to reach Red Lake, Roy St. John was sure he was near his destination. He dived down through the dense layer of storm clouds which buffeted and tossed the aircraft and when he broke through, he was flying close to the tree tops. There was no sign of Red Lake or any stretch of water — only forest. He climbed through the cloud layer again to fly further but there was no break in that swirling dark cloud. He descended again but his gaze met only a mass of tree tops. He leaned out the cockpit window to better see his way against the driving rain. It was no use. Although St. John knew the area he was flying over like the back of his hand, he was lost. He could not get his bearing as bush planes were not equipped with radios in those days.

Roy St. John decided to turn south and seek some body of water on which to make an emergency landing and wait out the storm. He again descended into that thin space between the storm and the endless stretch of swaying trees. Leaning out of the cockpit, straining against the blinding rain, St. John squinted through the dusk in an attempt to see shining water. Suddenly his wing tip clipped the top of a tree and the aircraft careened into a clump of tall pines. One wing tore off and it dropped nose down about 40 feet to the ground. Had it not been a Junkers aircraft of metal construction, it would have disintegrated and no one would have survived.

St. John was knocked unconscious. Carlyle, seated beside him in the cockpit, suffered only a sprained wrist. Wilson sustained a badly sprained knee and Drybrough suffered a broken leg, just above the knee. Carlyle got out of the cockpit and was able to pull out the unconscious pilot. He then helped Wilson to the ground while Drybrough remained in the cabin all night. He was able to get out in the morning with Carlyle's help. The same weather conditions prevailed the next day — thunderstorms, rain and hail. They were wet but not cold and they ate from the airplane's emergency supplies, supplemented by Charlie Wilson's buttermilk supply that he took along for his ulcer.

Wilson asked Carlyle to find an open space, to gather some dry wood and cover it with pine boughs and have everything ready to have it lit as soon as anyone heard a plane so it would create a smoke signal. Drybrough was certain that they would all perish before they were found but Wilson assured him that they would be rescued as soon as the weather cleared and planes came out looking for them.

The third day, September 3rd, the weather cleared and they heard a plane. Carlyle rushed to the pile of brush and set it on fire. In a little while, a plane swooped over the area, then disappeared. Drybrough at this point felt that the group would never come out of this accident alive as the plane did not circle the area but just disappeared. However, Wilson reassured him that they would be rescued as their whereabouts was now pinpointed. Nothing happened all day. Drybrough said that he appreciated Wilson's confidence, but where was the rescue party?

At about 5:30 p.m. they heard voices coming through the bush. It was a ground rescue party that included Dr. Wade of Ontario Hydro. Dr. Wade set Drybrough's leg and the three injured men were carried on stretchers through the bush for about two miles to the English River where they were put on a boat and taken to Ear Falls. They were then moved to Hudson and later to Winnipeg General Hospital.

Wilson related that he was never so glad to see Anton Ingard, Hydro's transmission line patrolman — a big powerful man of 6'4" who had no trouble putting the injured on stretchers and carrying them out with the help of the other Hydro personnel. All the injured recovered eventually and returned to their normal way of life.

Chapter Seven

Energy and Communication

The development of an area, uninhabited or otherwise, requires three main elements: transportation, energy and water resources, and communications.

The transportation industry — which we have already discussed in some detail — delivers equipment, materials and supplies for development of mines and towns. But these developments also require energy for their operations. Hence the necessity of hydroelectricity. This chapter will examine the development of a part of the Patricia District by Ontario Hydro and the Lake of the Woods Control Board (LWCB).

In the spring of 1928, construction work started on a regulating dam at Ear Falls at the outlet of Lac Seul. The dam was designed as a pier and stop log structure, 44 feet high with gravity end walls. Its overall length was 601 feet, with 20 sluiceways 14 feet wide. The potential storage basin thus created had a capacity of 3.3 million acres or 145 billion cubic feet. It was larger than the capacity reserve of either the Aswan Dam in Egypt or the Elephant Butte Dam in Texas. The dam also had a potential hydroelectric power capacity of 30,000 horse power under a head of 36 feet. The contract for dam construction was awarded to Morrow and Beatty Limited of Toronto. The cost was shared equally by the federal government and the provincial governments of Ontario and Manitoba.

The three governments' idea was that construction of the dam would benefit all through the power potential brought about by the regulation of the English River's flow. It was also a time when cheaper funds were available due to the Great Depression. In order to transport all the equipment and material to the dam site from the railhead at Hudson, every available boat and scow was put into use by such companies as the Hudson's Bay Company's transportation department, Red Lake Transportation and others. The Triangle Fish Company — precursor to the Patricia Transportation

Opposite page: The Dam at Ears Falls in 1934.

Company — was contracted to bring in 2,800 tons of cement in their 15-ton scows. They also moved the transformers since they had heavy scows capable of handling large equipment and high tonnage.

During the summer of 1928, Howey Gold Mines applied to the Ontario government for hydroelectric power. Arrangements were made to install a generating station at Ear Falls, Ontario as soon as the dam was completed, rather than going to the expense of building a second dam closer to Red Lake. Kart Biglow was awarded a contract to clear the right-of-way for a transmission line between Ear Falls and Red Lake.

The first generating station at Ear Falls was completed in December 1929. It first fed power into the system on Christmas morning to the delight of the townspeople and Howey Gold Mines management. In 1935, Red Lake Gold Shore Mines and McKenzie Red Lake Mines got on the power line.

To provide electric power to the Central Patricia and Pickle Crow Gold Mines, Ontario Hydro created a reservoir out of Lake St. Joseph on the Albany River by raising it nine feet. Lake St. Joseph had two natural outlets at its eastern end — Rat Rapids and Cedar Falls. The decision was made to have a generating station built at Rat Rapids. However, the dam and power house construction personnel were up against the problem of getting the equipment and materials delivered to the power site. The Root River Portage waterway was not completed so the thousands of bags of cement and other materials required for the job could not be handled over the route.

Due to freighting difficulties, Ontario Hydro engineers decided to use local materials

An 11 ton transformer is moved down a creek toward Root Portage.

Opposite page: The Howey Gold Mines at Red Lake, Ontario operated from 1930 to 1941.

The Honourable George Drew throws the switch and places a new generating unit in service at Ear Falls in 1948.

wherever possible. Hence, the 700-foot dam was built of rock-filled timbers made water-tight by steel sheeting on the upstream face. A four 164 RPM turbine was located in an open concrete flume on the downstream where the available 17 foot head of water would give a rated output from 1,000 to 1,200 horsepower. The turbine drove a horizontal 6,600 volt three phase 60 cycle generator.

During the winter of 1933-34, while Starratt's were hauling winter freight for Pickle Crow Mines from the Savant Lake railhead, Ontario Hydro had them transport the heavy equipment such as transformers, turbine and other bulky goods to the construction site. It amounted to approximately 3,652 tons. Food and lighter supplies were flown in. The generating station was built of logs, chinked with oakum and lined with one inch boards over building paper. The roof lumber was sawed on the job and covered with slate-surfaced rolled roofing. Four log cabins lined with insulating board were constructed, three to house the operators and one for the radio equipment.

Construction of the power plant was completed March 21, 1935 and the unit was brought into service. In 1936, a second unit was added to supply the four gold mines in the area over a 27-mile circuit. Ontario Hydro installed the second generator at Ear Falls in 1937. A 48-mile transmission line from Ear Falls to the Uchi Mine, along with an operator's house, was completed just before break-up in the spring of 1938.

During the summer of 1939, Ontario Hydro continued with the construction of their transmission line from Ear Falls to Pickle Crow. The pole line material that was shipped from Hudson to Root Portage was delivered by the Patricia Transportation Company to them at Bamaji Lake. This delivery was made over a system of lakes, rivers and short portages. Ernie Wright was in charge of this project while Slim Dart did the checking and billing of the material when it left Root Portage.

Ontario Hydro continued with this power line to Jason Mines at Casummit Lake and eventually connected with the Rat Rapids. This 113 mile extension from Uchi covered a total of 161 miles from Ear Falls and provided a source of power for other possible future mining developments. Five patrolmen's houses were constructed along the right-of-way from the Crow River transformer station to Uchi's switching station for the accommodation of the men patrolling the transmission line. These houses were set up at 20-mile intervals connected to Hydro's transmission line. Houses were located at Camp 12, Fry Lake, Harstone Lake, Kaw Lake and Uchi.

The patrolmen would not see any human beings for weeks on end. Their only connection with the outside world was the Hydro telephone. With the need for human contact, they would often listen in on the telephone and overhear any messages relayed on the line.

During 1939, a single-circuit 44,000 volt wood pole transmission line, 85.5 miles long, was built from the Ear Falls generating station to Sioux Lookout and Hudson. A distributing station was installed at Sioux Lookout and Hudson for the supply of power to these two communities. Sioux Lookout received electrical service August 30, while Hudson received it September 16, 1939.

The third generating unit was installed at Ear Falls in 1940, which increased the generator capacity of the station to 15,500 KVA. In 1948, the fourth unit at Ear Falls generating station was

Red Lake Fire

The town of Red Lake had a most disastrous fire when the three storey Red Lake Hotel burned to the ground on July 1, 1945. The fire started at 1:30 a.m.. Many guests who jumped out the windows suffered injuries, including Ernie Wright, general superintendent, and Don McLennan, treasurer, both of the Patricia Transportation Company, who happened to be there on business at the time. The seriously injured were flown out to Winnipeg during the night and the following morning.

Andrew Szura, 34, who worked as a dock hand in the summer and as a tractor driver during the winter for the Patricia Transportation Company, became a hero by saving three people. Seeing the building burning, he dashed into it and dragged one person out to safety. He went in again and dragged another person out and, to the horror of spectators, he again dashed into the inferno and dragged a third person out. After coming out the final time, his hair singed off and his whole body charred, he dropped unconscious on the street and was given first aid on the spot. In the very early morning he was flown to a Winnipeg hospital where he died from the severe burns.

There was no hospital at Red Lake at the time, so some of the injured and badly burned were rushed seven miles to the Madsen Red Lake Gold Mines hospital. Dr. Joe McCammon was the resident doctor who worked during the night and day and was credited with saving many lives. He later recalled, "There were some people so badly burnt that they were almost unrecognizable and there was nothing I could do for them."

installed and on July 27, Premier George Drew officially opened it. The premier tripped the switch which activated the rotor of the 5,500 kilowatt generator which increased the output at Ear Falls to 18,460 kilowatts, the 25,000 horsepower for which it had originally been designed 20 years earlier.

The local electrical service was extended the four miles into Goldpines and, for the first time the power generated at Ear Falls became part of a gigantic network of power stations throughout the province rather than a closed circuit supplying power only to local industry and towns. Due to the depletion of gold ore bodies, the Berens River Gold Mines closed their milling operations at Favourable Lake, Ontario in September, 1949. Ontario Hydro purchased their generating plant at North Wind Lake with a capacity of 1.5 megawatts.

With a demand for power downriver, Ontario Hydro decided to build a generating station at Manitou Falls, approximately 20 miles from Ear Falls. There was no road from Ear Falls to the proposed power site. Before a road could be built to the site, it was necessary to haul all the equipment and supplies by water from Ear Falls at the start of construction in 1954. To carry out this hauling operation, Ontario Hydro purchased the Patricia Transportation Company's old tugboat *Patricia* to move all the equipment and supplies by scows from the docks at Little Canada. However, instead of returning the boat to Little Canada where someone could have used it to good advantage on the Chukuni River, it was merely abandoned on

Rat Rapids Dam.

Camping Lake. It has since been salvaged and is now a proud part of the Ear Falls Museum.

During the construction period, Ontario Hydro built bunkhouses, a large cafeteria and a huge recreation hall for its personnel. They also built a 15-bed hospital with Dr. Truen in charge, with a staff of three nurses. Art Hollohan was the first aid man. Serious cases were flown to the Red Lake hospital.

Construction of the 72 megawatt Manitou Falls generating station was begun in 1954 with four units in service in 1956 and a fifth added in 1958. This generating plant was unique in that it was designed to be operated by remote control from Ear Falls, 20 miles upstream. The control link was by VHF radio and at the time it was said to be the largest radio-controlled power station in North America.

In 1958, an agreement was reached between the Manitoba and Ontario governments for the diversion of the headwaters of the Albany River into Lac Seul. A channel was cut along the Root River Portage route, which was part of the navigation route from the Hudson railhead to Doghole Bay during the years 1935 to 1953. A control dam was also built to regulate diversion flows and at the same time, the Rat Rapids power house was converted to a spillway.

When the level of Lac Seul rises above specified elevations, the Lake of the Woods Control Board (LWCB) controls the diverted flows. The rest of the time, Ontario Hydro operates the diversion to secure the most advantageous use of the diverted water for power generation in the two provinces.

Manitoba can request that the diversion dam be shut off when high flows occur on the Winnipeg River in Manitoba. When this occurs, the water can be stored in Lake St. Joseph for future diversion or can be released through the Rat Rapids Dam at the original outlet of Lake St. Joseph into Albany River and thence to James Bay. This diversion project increased the drainage basin of the Winnipeg River by nine per cent. These additional waters diverted through the Lake St. Joseph Dam now produce electricity by falling 660 feet rather than the original 17 feet at Rat Rapids on its journey to the sea.

Connecting to the Outside World

The advent of communications was also integral to the successful development of Northwestern Ontario. The first modern radio connection was achieved with the help of the Ontario Forestry Department's network of stations located throughout the Patricia District, with its base at Sioux Lookout.

By the early 1920's, the Forestry Department had radio stations located at strategic points in order to spot forest fires and report their locations to Sioux Lookout. Crews were then dispatched to the fires with surface craft.

On Lac Seul, they had a boat named *Wenesaga*, captained by Ole Sands, which operated on the 100 miles of Lac Seul between Hudson and Goldpines. Meanwhile, the *Chukuni*, piloted by Charlie Johnson, operated 65 miles in the Chukuni River waterway between Ear Falls and Red Lake. These boats transported men and equipment to the nearest scene of the forest fires. From these boats the firefighters proceeded further using outboard motors to get as close as possible to the fires.

With the introduction of aircraft, the department was able to expedite men and materials more quickly to the fires, while other boats followed through with additional men and equipment.

Originally, one of the radio stations was established at Pineridge (Goldpines), but due to the increase in the water level on Lac Seul by approximately 10 feet when the Ear Falls dam was built, it flooded out. So, the Forestry unit built a base on Goose Island, across the channel from Goldpines. In later years, it was re-established at Goldpines.

Opposite page:
The radio station at Kapikik Lake in 1929.

The other radio stations that were established in the Patricia District were located at Kapikik Lake, Pickle Lake, Red Lake and Woman Lake. During the winter the Kapikik and Woman Lake radio stations were closed while Pickle Lake and Red Lake remained open year-round as they were the only means by which mines could send messages to the base at Sioux Lookout. These messages were then turned over to the Canadian National Telegraph Co. agent for transmission.

The Goose Island radio station was open year-round not only to service Goldpines, but also during the winter it was of great help for the tractor swings hauling freight to Red Lake. Whenever an emergency arose or a tractor broke down, a crew member was sent to the station so a message could be sent to Hudson via Sioux Lookout for assistance.

The Red Lake station was originally established at Forestry Point; Sioux Lookout at the air base; Pickle Lake at Pickle Lake which was headquarters for that area; and, Kenora was at Kenora Lakeside, which was headquarters in that region.

The foregoing stations went commercial in a deal made with Canadian Marconi Company, which in turn sold the operation to the Canadian Pacific Railway. After the takeover, Marconi announced that installation of permanent radio-telephone connections throughout the main points in the district was being worked on. They also said that the system being prepared was to be one of the most modern and elaborate yet established for a public utility anywhere in the world at that time.

A headline in the *Red Lake News* on December 17, 1937 reported: "Radio-Telephone system for Patricia District now completed", and stated the happy news that Red Lake, Pickle Lake, Kenora, Sioux Lookout, Woman Lake and Winnipeg were soon to be joined.

Now it was clear that Northwestern Ontario was ready to move into the modern age with two crucial amenities–dependable power sources and up-to-date communications.

Company Characters

Dave Ross

Dave Ross and I were close friends in Winnipeg when the Great Depression started. We both found ourselves unemployed and we went our separate ways looking for work. He eventually got a job in Vancouver and I got work at Hudson. One day Dave wrote me and said that he was getting "web feet" from all the rain and he wondered if there was a chance of securing employment in the Hudson area. I told him to come right down as the office could use a man of his experience.

Dave arrived in the spring of 1937 and took up residence at the Grandview Hotel, which was situated on the ridge above the Hudson waterfront. It overlooked the south shore of Lost Lake, and certainly had a "grand view". All Dave had to do was walk out of the hotel, cross the tracks and he'd be in the company's office.

Dave was a city man and would not adapt to the northern style of clothing. He even wore city clothes all winter. However, when he got married and moved into a house, Dave ran into problems. During the subzero temperatures of winter, it was necessary to bucksaw enough firewood to maintain heat in the house, and Dave didn't like to get his hands dirty.

No matter how cold the weather, Dave would dash out of the house with woolen gloves and oxfords without overshoes, in order to bucksaw a couple of logs for firewood. I felt sorry for him and helped him by bucksawing his four-foot lengths of cordwood into stove-lengths so that all he had to do was to carry it into the storage space in the house.

Dave felt that we were both wasting too much time on firewood so he hit on a plan. Instead of three cuts resulting in four pieces 12 inches long, he would make two cuts, giving him three pieces, each 16 inches long. Sixteen-inch pieces were too long for the stove, so he instructed Mrs. Ross to put the piece into the stove and leave the door open. He also informed her that when the log burnt down sufficiently, she could shove it in further and close the stove door. Needless to say, the house was always full of smoke and you can just imagine what it did to the interior decorating.

Even while wearing his baseball sweater Dave Ross, kneeling far right, sported a tie. Also pictured in the front row are, left to right: Blackie Gunnison, Albert Pringle and Dick Roden. Standing are, left to right: unknown, Fred Wright, Ernie Wright, Albert Roden and Jack Wish.

Chapter Eight

Expansion Across the North

During the Second World War, the Patricia Transportation Company expanded its operations by purchasing all the ground-based assets of its old nemesis — Starratt Airways and Transportation.

As a result of this purchase, Arvid Anderson and I were assigned on December 31, 1941 to proceed to Ferland, 179 miles east of Hudson. Arvid was to take charge of freighting operations at Ferland and I was to take an inventory of the equipment and supplies we had just acquired.

It was about 30 degrees below zero when we boarded the Canadian National Transcontinental passenger train at 1:30 in the morning. Since it was about a five hour journey with stops in between, we decided not to take a sleeper in the Pullman car as, no sooner would we get bedded down comfortably, it would be time to get up, so we just dozed off in the passenger car seats. At about 6:30 a.m. on New Year's Day, we asked the train conductor if we had enough time to have breakfast before arrival at Ferland and he assured us that there was. We went into the dining car and were about halfway through our breakfast when the conductor came rushing in, waving his arms, and telling us that the train was not going to stop at Ferland. He shouted that the train was only going to slow down, and we would have to jump off.

During severe cold, steam locomotives have difficulty starting to move again after coming to a full stop. The train was already behind schedule, so the conductor told us we would have to jump. We dashed back to our passenger car seats, grabbed the duffel bags and eiderdowns and ran to the doorway of the passenger coach. When the train slowed down, we threw out our gear and leapt into the dark. This was about 7 a.m. and brother, was it cold!

We walked over to Starratt's camp office and introduced ourselves to the man in charge of operations, as well as to the billing clerk and freight checkers. He took us to the bunkhouse and assigned us a couple of bunks. After a proper breakfast, Arvid

Opposite Page: The Granduc cookery on the Salmon Glacier.

went out with the man in charge of operations to inspect the camp complex. I decided it was too cold to take an inventory, so I walked about three quarters of a mile to the shore of Lake Nipigon. It was now about 50 degrees below zero. The ice in the lake was cracking and it made a noise as though it was cannonading and you could also hear the bark on the trees cracking.

We had lunch at the cookery with the camp personnel after which we went out and looked over the camp site and its operations and discussed some of the problems with the various camp personnel. Supper was served at 6 p.m. after which everyone retired to their respective bunkhouses. It might be mentioned that in these northern camps, during the winter months, the niceties such as putting on pajamas and bedroom slippers were dispensed with. You just peeled off your outer garments and crawled into your sleeping bag with your underwear and a pair of woollen socks. Some chaps who were bald-headed put on a toque.

During the night, the "Bull Cook" went around the camp and stoked the fires in each bunkhouse and the other buildings. For some reason or other, whether by design to show his displeasure with the new company taking over the operations or, as an actual oversight, he neglected stoking the fire in our bunkhouse stove. In overlooking our stove, he did not endear himself to his fellow employees either, because they were also sleeping in the same bunkhouse. To make matters worse, the wind blew the bunkhouse door open. Nobody dared get out of their warm sleeping bags to close the door and stoke up the fire. Eventually Arvid Anderson jumped out of his sleeping bag, slammed the door shut and stoked the fire. During this time a torrent of unprintable words spewed forth. He crawled back into his sleeping bag until the bunkhouse was warm enough for us all to get up and get dressed. After this incident, the "Bull Cook" never overlooked carrying out his nightly duties of taking care of all stove fires in the camp.

In the north during the winter, the days are short and the work is mainly performed in the dark. Daylight usually begins about 9 a.m. with a misty blue haze, so anyone walking outside leaves an image of a ghostly body moving across the landscape. As the sun came up the temperature rose to about 35 below zero, but one still had to wear a parka hood over the head to help breathing. I went out with a clipboard to start taking the inventory and because it was so cold I had to wear mitts instead of gloves. It's difficult to write with mitts on, but it worked out satisfactorily because just as soon as I got too cold, I ran into the bunkhouse to warm up. While I was warming up I would try to decipher what I had written. I was accustomed to cold, but I had rarely encountered anything like those three days of severe cold at Ferland.

Northwest to the Yukon

After acquiring Starratt's, the Patricia Transporation Company further expanded by taking on a number of hauling contracts across the north. The founders of the company probably never dreamed that this small freighting outfit from a village in northern Ontario would one day be conducting long-distance hauling operations all across the Canadian north. But the Patricia company had a very good reputation for professionalism in long-distance freight hauling, and this reputation won the company contracts from as far away as British Columbia and the Yukon.

In 1955, the Canadian and American governments decided to have a mid-Canada radar line constructed across Canada. The Patricia Transportation Company was awarded the 550 mile hauling segment between Bird, Manitoba and Winisk, Ontario. The nearest railhead to the operation was Gillam, Manitoba and the Patricia Transportation Company's first job was to survey and select a suitable trail between Bird and Winisk for establishing Doppler radar sites. The route more or less straddled the 55th parallel of latitude. This trail was to be used the following winter freighting season for hauling equipment and supplies for the construction of these sites.

The Patricia Transportation Company received the contract for this project from Bell Telephone in early February of 1955. Upon receipt of the contract, the company immediately assembled the equipment necessary for the route survey from Thompson, Manitoba to the project at Thicket Portage, 150 miles southwest of Gillam. This equipment was loaded on the railway flat cars at Thicket Portage for Gillam and consisted of eight tractors (four D6's and four D7's), 40 tractor sleighs, six cabooses, one Bombardier, four 500 gallon tanks for the tractor swing's fuel supplies, as well as fuel oil for tractors, rations and other equipment and supplies necessary for carrying out the road survey and the following winter's freight hauling operations. Also, a dozen 1,000 gallon tanks were loaded for transporting customers' fuel oil for their storage tanks at the Doppler sites.

Jim Carson was in charge of the project and R.O. Turnbull was put in charge of the field operations. Carson was based at the Bell Telephone base of operations at The Pas, Manitoba and had an aircraft and pilot at his disposal to keep in touch with the ground crews. Assistant superintendent Turnbull was in charge of the ground operations. A crew of 30 men was also sent to Gillam, consisting of tractor drivers, brakemen, mechanics, caboose cooks and others for establishing a base camp and for the trail. Bell Telephone supplied the transportation company with a two-way radio and a supervisor to locate a Doppler radar site every 30 miles on the route from Bird to Winisk as the convoy of eight tractors with sleighs and cabooses went along. The Bell supervisor operated the radio and contacted his office at The Pas on a daily basis to report on the progress.

The first phase of operations covered the mapping out of a tractor route and the establishing of 18 Doppler sites on this 550 mile route during the last part of February and beginning of March in 1955. Bell Telephone also supplied a Norseman plane and a helicopter with pilots to be used by the company as required. Jim Carson, the general superintendent of the Patricia Transportation Company, said that they did well to map out the best tractor trail in one month's time. The use of the helicopter and Bombardier were the key to accomplishing this phase of operations. Unfortunately, one evening when the helicopter was coming in for a landing at the base camp, the tail section broke off and hit the main rotor blades. It crashed, killing the pilot and Elmer Davis of Woodlands, Manitoba, the company's spotter of the trail, and a veteran swing boss from the Lynn Lake project. The helicopter and crew had to be replaced.

Carson said that more men and equipment were taken in than was necessary, but at the time it was not known what problems would be encountered in mapping out this initial trail in an unknown, virgin territory. For example, Carson had been told that he would not be able to cross the Hayes River before spring break-up, so pumps were taken along so that the ice could be built up on the river to enable them to cross it with the tractors.

In the winter of 1964-1965, the Patricia Transportation Company was asked to carry out the hauling of fuel to Site 415, located about three miles from the west shore of James Bay and 150 miles east of the Winisk base complex. Instead of using two 1,000 gallon tanks per sleigh, which was

customary in hauling fuel oil in the early years, 2,000 gallon tanks were used on this resupply hauling operation. Bob Turnbull said that the last 40 miles of trail led over open rolling tundra. This open country could be lovely on a clear day but a fearsome hell in a blizzard.

Moving an Entire Town

One of our biggest projects entailed moving an entire town and mine site 165 miles across the Manitoba north. The Sherridon copper mine was situated approximately 41 miles northeast of Flin Flon, Manitoba, and was accessible by Canadian National Railway or by air from Winnipeg. When copper ore reserves at the Sherridon mine were being depleted, the Sherritt Gordon Mine's management decided that they would dismantle the plant and move it with the other mine's equipment, as well as the townsite, to their new nickel and copper property at Lynn Lake, 165 miles farther north.

With this decision, they hired Scotty Walker and Jim Carson from Ilford, Manitoba, who were in charge of winter freighting from Ilford to God's Lake Gold Mines, to organize their winter freighting operations. Scotty Walker was put in charge of winter transportation to Lynn Lake while the development and exploration was in progress and Jim Carson, master mechanic, was put in charge of tractor maintenance.

There was some miners' unrest at the Sherridon mine during the winter of 1946-1947, so Sherritt Gordon Mines were afraid that problems might develop, disrupting their winter freighting operations. They contacted Patricia Transportation Company to carry out freighting operations to their Lynn Lake property. This hauling contract came at a most opportune time for the Company as their 10 year winter freighting contract for the Berens River Mines was coming to an end.

Ernie Wright, Patricia's general superintendent, took over the Lynn Lake project, at which

Opposite page: House en route to Lynn Lake from Sherridon mine site.

Allan Sandin.

time Jim Carson was transferred to Patricia to be the Company's master mechanic and to assist Ernie with the management of operations. Carson's winter tractor freighting experience started in 1934 with Brooks Airways of Prince Albert, Saskatchewan. The company had the hauling contract to supply the Hudson's Bay Company's post north of Montreal Lake, Saskatchewan. Carson explained that the conditions were very tough, with many snowstorms and blizzards, along with temperatures dipping from 40 to 60 degrees below zero. It took three months to make the round trip.

A base camp complex was set up by Patricia Transportation at the Sherridon Mine property, consisting of a garage to take care of servicing and repairing of tractors, a blacksmith shop for repairing sleighs and racks, a bunkhouse for its personnel, as well as a cookery to take care of providing meals to its employees. Incidentally, these facilities were kept open throughout the year, as after each freighting season the mechanics remained to overhaul the tractors and the blacksmiths remained to repair any sleighs damaged during the hauling season.

Wright brought in veteran drivers such as Mel Parker, Jimmy King, Sig Erickson, the Vincent brothers—Laurie, Lester, Stewart and Wilfred—and some caboose cooks, including Tim McGinnis who was with the company for many years. Other drivers were recruited from farms in Manitoba and Saskatchewan.

Employee Allan Sandin, who was in charge of the unloading operations at the mine's "Laurie River" power site, recalled that the lowest temperature reading at the site was 63 degrees below zero, and that when it got up to 40 below, they thought it was a mild spell. Their garage was a tent, so the tractor had to run continuously. They dared not stop it for servicing as they were afraid it would not start again. The motor was kept running until a swing arrived from Sherridon when it was stopped and serviced and restarted again. Sandin's camp had a short wave radio to communicate with Sherridon and planes flying in the area. He said communications weren't very reliable during the northern lights, due to the electrical interference.

A tractor with driver was kept at each base location to carry out the spotting of sleighs. The reason this was necessary was the fact that the swings did not stop long enough to carry out the unloading. When a swing arrived at a destination, it dropped off its loaded sleighs, picked up the empties and was on its way back to Sherridon to pick up another loaded swing of sleighs. Time was of the essence, as winter freighting in that part of the north only lasted approximately 90 days per winter season and rarely did it last 100 days, so swings operated around the clock regardless of the weather, Sundays or holidays.

Winter freighting season in the area usually started in the early part of December when the trail breaking and checking of ice conditions was carried out. A Bombardier was best suited for this type of work as it had long rubber tracks with cleats. It could go over a minimum of two inches of ice. It also packed the snow down on the muskeg to let the frost through and freeze the ice deeper in order to carry the heavier weight of the tractors. A road maintenance crew was also maintained, utilizing one tractor driver, three helpers and a caboose cook. Their work consisted primarily of maintaining the portage approaches on the route from Sherridon to Lynn Lake.

One year's supply of wood — 5,000 cords — at Lynn Lake.

A variety of materials and equipment were hauled into the mine. Lumber and timber from the mine's sawmill at Granville Lake were hauled in, while ore was hauled back out of the mine. Jimmy King, one of the tractor swing bosses, said that during one winter season they had to pick up a load of fish on their return trip from Granville Lake. The fish, including trout and pickerel, was in huge piles on the ice and was just thrown loose onto the sleighs. It was dropped off at the mine's Cold Lake camp on Kississing Lake, about 30 miles north of Sherridon.

The first house was moved from Sherridon to Lynn Lake in December 1951. The average size of house moved was 36' x 24'. The only thing braced was the chimney. Special sleighs were built for moving buildings and only one house was hauled per swing. In one of the company's more remarkable operations, Patricia Transportation eventually moved a total of 219 buildings from Sherridon to Lynn Lake, including a bank, church, hospital and a two-storey school house.

Woodcutting Operations

Clifford Snowdy of Sherritt Gordon Mines was put in charge of cordwood cutting for the Lynn Lake mine property. During three years, 15,329 cords of wood were cut in four-foot lengths.

The woodcutting operations were carried out within a 25-mile radius of bush around the mine property. This cordwood was used during the development stages of the mine for steam boilers and the like, as well as for heating purposes. Every year there seemed to be a large discrepancy between the wood cut in the bush and that actually delivered to the mine yards. Sherritt Gordon Gold Mines asked the Patricia Transportation Company if they could arrange to have one of their men supervise the woodcutting operations since the company was already hauling the cordwood from the bush to their yard.

Harry Everett was selected by Patricia to be in charge of this operation as he was adept in handling native workers and had considerable experience in open country operations, as well as being an educated man. On September 15, 1950, the company asked him to leave Doghole Bay, Ontario, where he was the company's agent, to proceed to Lynn Lake to be in charge of the mine's wood cutting project. Upon arrival at the Lynn Lake mine, he was assigned quarters in the mine's staff house and he was to take meals at their cookery when in camp.

Allan E. Gallie, the mine manager, told Everett that the annual cutting target was 5,000 cords of wood. The bush country within a 25-mile radius of the mine was not a very good cordwood cutting area. The best timber stands were located along the shoreline of various nearby lakes. Normally it took about 16 trees to provide a cord of wood. Around Lynn Lake, it took about 40 trees. This was due to the fact that, this far north, the trees are fairly thick at the bottom but they are not very tall. In spite of the poor timber area, the woodcutters were able to average 30 cords per month.

Everett patrolled the area to locate the best timber stands and sometimes climbed the tallest tree to spot the good sites. He scaled and checked the cordwood and paid the cutters at cutting sites. There were 10 camps in operation, consisting of two freight cabooses on skids per camp. There were anywhere from 10 to 30 woodcutters. However, you never knew if you would find the cutters in the camps as some of them would just take off, leaving everything behind.

Woodcutting operations were carried out year-round. In the summer, if Sherritt's planes were busy, Everett had to walk or paddle a canoe and portage it. He also had to pack the cutters' groceries and deliver them. He said that the lakes were very pretty along the shore line after the leaves had turned different colours in the fall. Everett also noted that the mosquitoes and blackflies were very bad during the summer. There was no problem at night as he slept under a mosquito net, but it was quite a different story during the days. He said that one day he went down the Keewatin River to check on cutters at Cocherman Lake and deliver some groceries. At night time he camped by the river and the next morning he tried to cook breakfast but was unable to do so as there were too many mosquitoes. After freeze-up, before the Bombardier could be used, he had to pull a hand sleigh with his sleeping bag and groceries for the cutters. Later on, when the ice was safe to travel on, he used a Bombardier to check the camps every day.

The cordwood was hauled to the mine yard by D6's and TD14's as well as Linn tractors with two or three sleighs with about four or five cords per sleigh, depending on conditions. The D6's and TD14's were the best suited for hauling as the Linn tractors travelled too fast and often lost lengths of wood while en route from the bush to the mine yard.

During the seven years of cordwood cutting operations, the company hauled 35,329 cords of wood from the bush to the mine yard, of which 15,329 cords were handled under the supervision of Snowdy, the mine representative and the balance of 20,000 cords were under the supervision of Harry Everett of the Patricia Transportation Company and Lester Vincent, a long-time employee of the company who looked after the loading of the cordwood in the bush. Sherritt Gordon Gold Mines complimented Harry on the excellent job he had done for them.

Working in the north is often dangerous, and there were accidents on the woodcutting operation. Two men died at Bighetty's camp. When one of the cabooses was being skidded by a tractor to a new cutting site early one morning, the other caboose was left behind. One of the four remaining men was going to start a fire in the stove to make breakfast. The light in the caboose is not the best at any time, so when the man picked up a can which he thought was kerosene to help get the fire going better, it unfortunately was a can of gasoline, which exploded and set the caboose on fire. The men's clothes were burnt, and they barely escaped from the caboose with their lives.

Two men were badly burnt and the other two decided to walk to the Lynn Lake mine for help. There was some underwear hanging on the line, so each put an extra pair of underwear on, tied rags to their feet with building paper wrapped around their legs and started to walk facing the wind at 40 below zero. One of them walked 12 miles and was within two miles of Lynn Lake when he met a tractor on the way to Lynn Lake from Sherridon, which picked him up. He passed out and was brought to the mine, then put on a plane and flown to the Sherridon hospital. On the way out, the pilot spot-

Harry Everett and the Bombardier he used on the cordwood cutting operations at Lynn Lake.

ted the other man walking on the lake, so he landed beside him, picked him up and took him to the Sherridon hospital as well. Everett had the task of going with the Bombardier to the burnt out camp site to pick up the two badly burnt men, but by the time Everett arrived they were dead.

Another accident occurred when Ernie Wright, general superintendent of the Patricia Transportation Company, came up to Thicket Portage with his brown wooden Bombardier from Wekusko to check the road to the Moak Lake mine property. The next morning, he left to go and meet the tractor swing. Since Ernie was going right to the mine at Moak Lake, Harry Everett got him to take three quarters of frozen beef for the mine cookery. He was about 15 miles from Thicket Portage on Paint Lake when the Bombardier caught fire. Ernie and his two passengers were able to get out before the gas tank exploded, but were unable to get the meat out and there were three quarters of beef well roasted, lying on the ice. Not long after the accident the tractor swing from Moak Lake picked the men up and brought them back to Thicket Portage.

Ernie felt very bad about the loss of his Bombardier, which he had fixed up very nicely and in which he had travelled many miles throughout the years during the winter seasons. One year he went from Churchill, Manitoba to Rankin Inlet in the North West Territories, a distance of approximately 400 miles.

In November 1954, Harry Everett was sent to Thicket Portage, Manitoba, located on the CNR Line, to be the Patricia Company's agent and supervisor of the winter freighting operations. During that same winter the mine at Moak Lake required some gravel and sand, so they sent some miners to Thicket Portage where there was abundant gravel. The company hauled about 12 loads of it to the mine property. In the spring, the snow melted and the mine people found all kinds of gravel and sand in a hill near the mine. Had the mine people known of this gravel site, they would not have gone to the expense of hauling it 65 miles from Thicket Portage. It was bad luck for the mine but good luck for the Patricia Company, as we made some extra revenue from the haul.

Paving the Way for a New Boomtown

In 1956, when Inco announced the major discovery of a nickel-copper deposit at Thompson, Manitoba work on the earlier discoveries at Moak Lake was suspended since a higher grade of ore was found at Thompson. The site of a proposed townsite at Moak Lake was moved to where the city of Thompson is now located. The change of the mine site meant all the equipment and supplies that were delivered to the Moak Lake property during 1954-1955 and 1955-1956 winter freighting seasons had

Thompson Townsite

The Patricia Transportation Company arranged for the clearing of the bush for the Thompson townsite. Although it is hard to believe today, the stretch of bush that was just uninhabited wilderness in 1954 is now Manitoba's third largest city with a population of approximately 15,000. It has become the world's second largest nickel producing centre.

to be hauled back to the new mining site at Thompson. This was much-welcomed revenue for the Patricia Transportation Company.

Due to the large tonnages to be handled, Jim Carson was put in charge of this project, with Harry Everett as agent and biller of equipment, materials and supplies. A base camp with all the facilities was established at Thicket Portage. Since this was a short haul, they were able to load the sleighs with as much as 10 tons, depending on the type of equipment or material to be hauled. Throughout the three winter freighting seasons of 1954-1955, 1955-1956 and 1956-1957, thousands of tons of equipment, materials and supplies were hauled from the railhead at Thicket Portage to the Thompson property, a distance of approximately 40 tractor route miles. Thousands of feet of lumber and thousands of bags of cement were delivered the first two seasons. Assorted other heavy equipment was also delivered, such as steam boilers, hoist equipment, and diesel engines.

The mine had to have thousands and thousands of gallons of fuel oil for diesel engines in order to carry out the construction program, so steel storage tanks were brought in sections and hauled from the railhead to the mine where they were assembled on the mine's tank storage area. One of these storage tanks had a capacity of 75,000 gallons and there were another 20 storage tanks with a capacity of 19,000 gallons each. The last year of winter hauling operations, one million gallons of fuel oil and gasoline were delivered to the mine's tank yard. The mine had to have this volume on hand for their development operations until the railway spur to it was completed from Sipiwesk, located on the Canadian National Railway's Hudson Bay Line. When the railway spur was completed to Thompson in 1957, the Imperial Oil Company set up their own storage tanks at the spur siding to meet the mine's requirements. The mine then dismantled their storage tanks and the sections were shipped out.

Rescuing a Cat

The Midwest Drilling Company wanted to do some diamond drilling for copper 60 miles west of Bird, Manitoba. They rented a D6 tractor, sleigh and caboose from the Patricia Transportation Company and hauled their equipment to the drilling site.

It was early in March of 1958 when they finished and the ice was deteriorating. On their way back to Bird, the D6 broke through the ice and sank in 10 feet of water. I was in Winnipeg at the time and had to fly up there and attempt to salvage it fast or it would be there for the summer. We loaded another D6 on a railway flat car at Gillam, along with a sleigh and caboose, as well as other equipment and food supplies that I felt were needed and shipped it all to Bird.

Upon the equipment's arrival at Bird, it was unloaded and I and one other man proceeded with our D6 tractor, sleigh and caboose to the site of the sunken tractor. When we were within 10 miles of our objective, our D6 went through the ice on the edge of a lake (the ice is always weaker near the shore from the heat of the land). We had a radio with us so I contacted Midwest and asked them to fly out three block and tackles with 200 feet of cable. There was a lake nearby where the plane could land. They also sent us two toboggans for us to haul the equipment to the salvage site.

It was muskeg country and there was only about three inches of ice (or frozen muskeg) on the surface, which did not make a very strong anchor for our block and cables. However, we made two anchors for each and managed to drag the tractor out. We were then able to continue on and salvage the other tractor. The other tractor had gone down below the water level and sucked in water through the intake and bent one of the connecting rods (water will not compress like air and this always happens if the motor is running). We were lucky as the motor would still turn over and run, although of course it was not firing on that cylinder.

There was no insurance on the two D6's and the cost of salvaging them the next winter would have run into a considerable sum of money and probably some additional costs from broken castings on the one that was only half submerged due to frost damage. I don't know who was the happiest, Midwest or Patricia, when we arrived back at Bird with both tractors. They had assumed that the situation was hopeless. And I should have been happy myself, but I was too numb and tired to have any feelings left.

One of the more imposing tasks was to haul two 50-ton draglines from the railhead to the mine property. They had to put these on 12"x12" timber skids pulled by two tractors. A rough lock was put on the rear of each timber skid going downhill with two tractors at the rear holding back the load to relieve the pressure of going downhill so as to eliminate the danger of jack-knifing. They also had to pump water on river crossings to build the ice up to at least four feet in order to haul the huge draglines across.

During those three winter hauling seasons, three crawler tractors broke through the ice with their drivers. On one occasion, the driver bobbed up and was rescued. On another occasion on Wintering Lake, the hole became plugged with the sleigh load and the driver, Jack Sutherland, drowned. On the third occasion, Barry Henry of Grandview, Manitoba drowned while crossing the Burntwood River.

After waiting about ten days for the ice to get stronger in the area of the accident, salvage crews set up tripods over the site and salvaged each tractor. Barry Henry's body was not recovered until the spring.

The Patricia Transportation Company also arranged for the clearing of a 60-mile transmission line from Thompson to the mine's hydro electric generating plant at Kelsey, located on the Nelson River. The company hauled the necessary materials to the sites of the towers during the winter of 1959-1960. The line followed the Burntwood River most of the way so boats were used in the summer for transportation.

The following summer Jim Riley was put in charge of construction with Jim Carson as supervisor. Eighty-four steel towers were erected and the wires were strung the same summer. No tractor hauling was necessary to Inco's hydro-generating plant at Kelsey as the Canadian National Railway built a 10-mile spur from Pit Siding on their Hudson Bay line to the power site over which all the equipment and materials were hauled by rail.

The Menace of Mountain Hauling

The Granduc property is located in the Coast Range Mountains of northern British Columbia near the south end of the Alaskan Panhandle and at the headwater of the Leduc and Unuk Rivers. Considerable prospecting took place there during the 1930's. However, access to the mountainous Unuk terrain proved exceptionally difficult because of numerous glaciers and the lack of feed for pack animals.

In 1951, the Granduc outcrop of copper was located by prospectors employed by the Helicopter Exploration Company Limited who in 1952 optioned the 207 claims to Granby Consolidated Smelting and Power Company Limited. In 1953, Granby incorporated Granduc Mines Limited with a capital stock of 4,000,000 shares of $1 par. In December 1953, Newmont Mining Corporation joined with Granby Consolidated Mining Smelting and Power Company in providing funds for the development of the property.

Stewart, British Columbia was selected to be the mine office with a two-way radio communication system between the mine office and the camp site located northwest of Stewart. Stewart is a small town at tidewater on the Canadian side of the Alaskan border at the head of the Portland Canal. It is reached by ocean vessels or by air 120 miles from Prince Rupert or 80 airmiles from Ketchikan, Alaska. The Granduc property and Stewart are separated from each other by 25 airmiles across ice-filled passes more than 5,000 feet above sea level.

In January 1954, the first airborne supplies were transported to the camp site from Stewart. Later, some supplies were moved by tractor train by way of Salmon and Leduc Glaciers and connecting snowfield, which at one point along the route was 6,000 feet above sea level. In November 1954, a short airstrip was built in the Leduc Valley, three

Snowcat at the Granduc project in 1955.

miles from the mine site. In winter and spring, the mining company's Super Cub operated on skis by landing on the snow-covered glacier. In the summer, an alternative airstrip was used. The first camp was established on the north side of Leduc Glacier on January 25, 1954. The camp had to be moved twice a year because of the danger of snowslides. In November it was located on the glacier for the winter months and in June it was moved back on the hill near the 3,250 foot level portal.

The problem of transporting heavy equipment to a property like Granduc is that it is situated 4,500 feet above sea level — in a region of changeable and often severe climatic conditions — surrounded on all sides by rugged mountains, glaciers and snowfields. The annual snowfall in the region is 800 inches or approximately 67 feet, with snowpacks between 20 and 50 feet. For the miners to carry out further exploration and development work, it was necessary to haul in large tonnages of heavy equipment such as diesel engines, compressors and the like. The mine and other companies in the area were not capable of doing this.

The mine company had attempted to survey a proposed access road up the Unuk River Valley, but with no success. During 1956, an attempt was made to test the thickness of the ice on the glaciers for possible hauling routes. Nine holes totalling 13,297 feet were drilled with special electrical hot point drilling equipment on the Salmon and Leduc glaciers. One hole in the centre of the glacier penetrated 2,365 feet of ice. One hole was drilled in the snowfield at the 5,000 foot level elevation between the head of the Leduc Glacier, but remained unfinished at 770 feet.

In order to carry out the hauling of needed supplies, the mining company contacted the Patricia Transportation Company to do the job, as it was the only qualified winter tractor freighting company in Canada with expertise to handle such a project.

Ernie Wright, Patricia's general superintendent, was put in charge of the Granduc project. In the fall of 1953, three D6's and four D8's Caterpillar tractors, along with two muskeg tractors and a Tucker Snow Cat, three cabooses and 20 sets of tractor sleighs were loaded on Canadian National Railway flatcars at Thicket Portage, Manitoba. As well, a supply of fuel oil for the tractors and other necessary equipment to establish a base camp for winter freight hauling operations were shipped to Vancouver, then by barge through the Portland Canal to Stewart. Food supplies were shipped from Vancouver.

Supply tents at a northern camp.

In early December 1953, Wright, with a veteran 15-man winter tractor freighting crew, flew to Stewart to establish the base camp and begin the freight hauling from Stewart to the Granduc Mine property. The freight was hauled by trucks for a distance of 11 miles from Stewart to a point near the foot of the Salmon Glacier and then transferred to sleighs and hauled by tractor over the high glaciers and snowfields to the mine property, a distance of 24 miles.

The mine provided a prefabricated building for a bunkhouse. There was also a cookery with a cook at the halfway point for the tractor crews to have their meals or stay at the bunkhouse if they got caught in a severe snowstorm and had to wait it out. Hence, no caboose was hauled by each swing.

The tractor drivers soon discovered that winter hauling operations were going to be entirely different to what they had to cope with on other projects in the rest of the country. Conditions were so hazardous that the freight hauling was only carried out during daylight hours. It took two days to make a round trip from Stewart to the mine and return, although sometimes it took as long as three or four days. Although three swings were in operation, they could only haul one loaded sleigh per tractor instead of the customary three or more sleigh loads as on other hauling projects. It was rare that a tractor could haul more than two sleigh loads, depending on the type of freight handled.

One can just visualize the hazardous conditions that were encountered when a monstrous 140 HP D8 tractor weighing 70,000 pounds or 35 tons could pull only one sleigh load of freight weighing from six to 10 tons. Each tractor was equipped with a winch at the rear so they could winch each other out when they got stuck or slid off the trail. These tractors were also equipped with bulldozers. Evert Cummer, one of the veteran swing bosses, said that they did not experience any jack-knifing accidents or need to "double up" as only one sleigh load per tractor was hauled. However, on numerous occasions, a tractor with its sleigh load of freight would slide off the beaten trail on side slopes of 35 degrees or more and would have to be winched back up onto the trail.

Cummer said that numerous cracks in the glaciers were encountered. They had to pack them down with snow, then bridge them with 12" by 12" timbers to enable the tractor sleigh load to cross.

All of Patricia's hauling expertise was needed to cross this difficult terrain at the Granduc project.

He said that fortunately on his swing some of the cracks that the tractors slid into were only five or six feet deep and the machines could be winched out. However, he said that some cracks were so deep that one could not see the bottom.

Cummer also related an incident during one of his trips when a swing of three tractors and three sleighs ran into a very severe snowstorm. They could hardly see in front of the machines and one of the tractors slid off the trail with its sleigh load. They could not stop to winch it back on the trail, so when its driver scrambled up to one of the tractors, they carried on to the mine as otherwise, if they had stopped to try the winch the tractor out, they might have been weathered in and perished. They were "stormstayed" at the camp for several days. When the storm subsided, they went back to where the tractor had slid off the trail and all they could see was the top of the tractor's exhaust pipe. This meant that the snow was approximately eight feet deep surrounding the tractor. Two bulldozers had to be used to pull the tractor from the packed snow.

In spite of the hauling hardships over a rough terrain at high altitudes and with severe weather conditions, the Patricia Transportation Company crews delivered approximately 1,000 tons of freight from Stewart to the mine property. This was hauled between December 1953 and March 1954. In 1955, Granduc Mines again hired Patricia Transportation Company to haul additional heavy equipment and supplies to the mine, and 2,096 tons were hauled between February 3 and April 20. During the winter months of 1955-56, the company transported an additional 2,011 tons.

Black Gold in the Yukon

In the summer of 1962, the White Pass and Yukon Route Company won the contract to set up drilling facilities for Standard Oil at the confluence of the Blackstone and Peel rivers, about 60 miles below the Arctic Circle in the Yukon Territory.

The oil company hired Patricia Transportation Company to haul the drilling rig as well as other equipment and supplies during the winter of 1962-1963. In order to carry out this hauling project, the company loaded onto Canadian National Railway flatcars at Thicket Portage, Manitoba (the base of their Thompson projection operations) three TD18 tractors, 14 sleighs, three cabooses, fuel oil for the tractors, as well as other equipment and supplies to last during the freighting season. Also, sufficient food supplies were shipped to last the season.

After the equipment and supplies arrived by rail in Vancouver, British Columbia, it was loaded on a boat to Skagway and then taken by a narrow gauge railway to Whitehorse, Yukon. There it was loaded onto trucks and hauled to the staging area base camp at Chapman Lake, which was 50 miles from Dawson, Yukon, near the Alaska border. This base camp was about 400 miles south of Inuvik located at the mouth of the McKenzie River on the Arctic Ocean.

Jim Carson was in charge of this project with R.O. Turnbull as the swing boss. Early in December they flew into Chapman Lake from Winnipeg, along with the tractor hauling crew and other personnel. It was unusual to have a woman cook on a winter tractor hauling operation, but this time Mrs. Carson came along to be the cook at the Chapman Lake base camp.

Upon arrival at Chapman Lake, Bob Turnbull, with a crew of five men, including the caboose cook, set out with a TD18 tractor and one sleigh of fuel oil and other rigging and supplies, as well as the caboose, to map out the 110 mile trail to the Standard Oil Company's wellsite. In the meantime, the balance of the personnel set up the base camp facilities for the winter freighting operations.

Freight-hauling started late in December of 1962, and Bob Turnbull said that it was a very rough route. The trail proceeded through canyons with high mountains on each side. There were flat areas of about a quarter mile on the Blackstone River. However, in spite of the extreme cold weather, when temperature sometimes dropped to 78 degrees below zero Fahrenheit, the Blackstone River flooded these flat areas to a depth of three or four feet. It was necessary to cross the river 17 times en route to the wellsite, so on occasions there was much wallowing in frigid water. Turnbull said it was rather scary for the drivers.

The crew consisted of four tractor drivers, which included the swing boss, two brakeman and a cook, or a total of seven men. The hauling operations were carried out in the usual work shifts of eight hours beginning with 12 noon, 8 p.m. and 4 a.m. every day regardless of weather conditions. The only time they stopped was to change shifts and to refuel the machines. The crew coming on shift would eat while the crew coming off shift would refuel the machines. Then the retiring crew would eat and sleep until it was their turn to come on shift again.

In the middle of the trip the temperature was 54 degrees below zero and they did not see the sun for six weeks. This area of the north experiences about six months of darkness during

In 1950, the Patricia Transportation Company redirected equipment away from work in the north country when scows were sent to the waterlogged region of southern Manitoba. Several scows were used to move livestock to higher ground during the worst flood in the modern history of Manitoba.

the winter months and almost six months continual daylight during the summer.

The freight on the sleighs included the drilling rig, power plant, a DC3 tractor, truck, prefab buildings, fuel oil storage tanks and fuel oil. A considerable quantity of drilling mud material was hauled for the drilling of a deep wildcat well. The drilling muds consisted of Barite, Bentonite and Bringel, as well as chemicals such as caustic soda, Dricose, phosphates, Querbache, and soda ash. The company also hauled the necessary materials and equipment for a gravel airstrip which they intended to build two miles from the wellsite. To move personnel back and forth from the airstrip, the oil company provided a Nordell Tracked Carrier.

During the winters, you could hear the roar of the tractors coming for miles in the cold as the temperatures dipped to 40 or 50 degrees below zero. Once they arrived at the drilling site, the work crews began construction of the camp. Once the airstrip was built, the fastest way out to civilization was by Standard Oil Company's executive plane, C-46. Needless to say, it was more posh than the caboose used by the Patricia Transportation Company's freighting crews.

For building the drilling rig, the piles were deliberately frozen into the permafrost. The piles were capped three feet above ground and timber joists for the drilling platform were laid. Since most of the area was muskeg, the peat moss was scraped from the surface and piled on top of the permanently frozen surface between the joists to provide insulation and keep the surface from melting in the summer, when temperatures sometimes rose to 90 degrees Fahrenheit.

Storage tanks were set up for the storing of fuel oil for the operation of the drilling rig as well as other diesel operating equipment. Prefab huts were set up to accommodate four men each of the oil company's crew. A prefab office building was set up, as well as a radio shack, power shack and a "doghouse". The latter was a small building in which the geologist checked the drilling mud samples and where the drilling mud engineers kept their drilling records.

The crew also constructed a good modern kitchen and mess hall that seated about 20 men. The washroom building boasted hot showers and all the conveniences, but the water was hauled in by trucks from five miles away and sometimes people ran out of water in the middle of a shower when the truck froze up.

Drilling of the well began in February 1963 and it was drilled to a depth of 10,000 feet during the summer. Unfortunately, it turned out to be a dry hole. It was a big disappointment after all that work, but that's the nature of the oil business. The drilling rig and other equipment were loaded back onto tractor swings and taken back to the Chapman Lake base camp, where it was sold at a low price to the White Pass and Yukon Railway Company. The price was rock bottom simply because it was too expensive to ship it back via truck to Whitehorse, by railway to Skagway, by boat to Vancouver and then by railway to Winnipeg.

End of an Era

Many of the jobs undertaken by the Patricia Transportation Company were unprecedented in Canadian history. The clearing of the Thompson town site and the subsequent long-distance freight haul into that area in 1957 were mammoth projects that simply have no comparison in the history of the North. But the "resource boom" soon began to level off. The gradual slowdown of the hauling business in the early 1960's was not because the company was doing a poor job. Indeed, one of the greatest ironies affecting the Patricia Transportation Company was that it did its job almost too well.

According to George T. Richardson, the current Chairman and Managing Director of James Richardson & Sons, Limited which owned Patricia originally, the company required a regular supply of new projects to keep it busy. "Unless someone was opening a new mine or hydro site in a remote area, we had no business. By the late 1950's, the bulk of the northern mines had already been established, so we began shifting our focus as a hauling company on southern business."

In its final years, the Patricia Transportation Company became a contracting and trucking firm that hauled freight across Western Canada and Northwestern Ontario. The contracting division remained active, clearing numerous hydro line rights-of-way through places like Grand Rapids, Leaf Rapids and Kelsey. The Patricia's construction crews also built much of the Trans-Canada Highway east of Winnipeg, the same road over which the Company's trucks hauled freight on a nightly run to Thunder Bay.

The Patricia Transportation Company was quite successful in the trucking business. But again, its tendency to complete the job in a timely and efficient manner meant that it did not last long in the competitive world of long-haul trucking. In 1956, a British firm came to Canada, shopping for trucking companies, and the purchasing agents took a liking to the Patricia's new IBM computerized accounting system. "It was quite revolutionary at the time," says George Richardson, "and they decided that they needed to have it. We weren't very interested in selling the company, so I didn't encourage these gentlemen. But they were persistent, so I wrote a number on a piece of paper and held it up. 'That's the price,' I said. I thought they would be dissuaded. But within half an hour, they were back with a cheque."

As Richardson recalls that day, there is a note of regret in his voice. The Patricia Transportation Company was always a favoured business, and selling it was a difficult decision, even at a top-of-the-market price tag. But the fact was, the company no longer existed anyway; at least not in its early form — as a loose-knit organization of pioneers and misfits and heroes who delivered millions of tons of freight in impossible conditions. Not even the postal service can claim to have braved the "rail, hail, sleet and snow" faced by the men of the Patricia. And it's a tribute to the decency of the managers that the company accomplished all those dangerous jobs in such a sensible manner. Always, safety came first.

The Patricia Transportation Company was more of a family than a business. In that sense, perhaps it is fitting that the company passed away at the dawn of the computer age — an age that would also see the passing of many of the pioneering values that created the company in the first place. "To tell you the truth, we never made much money from the Patricia Transportation Company," says George Richardson. "But we stayed behind it, because we felt the company was doing something important."

THE PATRICIA TRANSPORTATION COMPANY 1931-1964

It all began in 1920, when three men — Ole Gustafson, Carl Johnson and Ole Sands — decided to get into the fishing business on Lac Seul. They named their venture the Triangle Fish Company. Gustafson, being a Swedish boat builder, immediately started to construct a 48-foot fishing boat which was named *Triangle*. Its beam was 18 feet and it was powered by a three cylinder Calenberg semi-diesel engine. The company carried on with its fishing business until the first gold fever hit the area in 1922. The three men then sold the company and went gold prospecting in Red Lake.

In 1925, Gustafson joined up with A.S. Brown of Kingsville, Ontario. Together, they launched a business centred around exporting and trading in fish products. They raised $20,000 in capital, of which $6,667 was Gustafson's share and the balance was Brown's. The loan was secured by boats like the *Archibald* and the *Triangle*. On December 3, 1925, Wilfred Wright bought out Arthur Brown's interest and entered into a partnership with Gustafson on a 50-50 basis, with Wright as general manager and Gustafson as president. They continued in the fishing business.

Due to the 1926 Red Lake gold rush and the construction of the Ear Falls dam in 1928, there was a great demand for the hauling of equipment and supplies. The only tugboats and scows available to handle such a huge volume of freight belonged to the Triangle Fish Company. Seeing the great potential of a transportation service from Hudson to the northern mines, Wright and Gustafson decided to go into the freighting business in a big way. So they changed the company name to Triangle Transportation Company and acquired additional capital for expansion.

On September 21, 1931, the company expanded again and changed its name to the Patricia Transportation Company Limited. It had an authorized capital of 50,000 shares of $1 each, of which 30,005 shares were subscribed and fully paid up for the purchase of the transportation assets of

Opposite page: Staff attending a banquet at Winnipeg's Marlborough Hotel after completing winter operations at Berens River in April 1939. From left to right Back Row: J. McLaren, Allan Haycock, D.E. Wilson, A. Sandin, Oscar Evans, D. Learmonth, M. Vogan, R.J. Stutt, M. Matty, J. VanLancher and H. Simmons. Second Row: G. Sayers, Arvid Anderson, D. McLean, D. Bush, S. Erickson, C.L. Fisher, T. Toweys, T. Fisher, Norman Howard, E. Snowball, Carl Robinson and T. Jones. First row: Alex Anderson, Gordon Lawson, P. Timchuck, Ernie Wright, C.T. Wilson, E. Baumgaard, E. Emms, A.E. Roden and R.A. Booker.

the Triangle Transportation Company and to conduct the business of passenger and freight hauling from any point in northern Ontario. Company headquarters was Hudson.

The major shareholders in the new company were Ole Gustafson and Wilfred Wright. The holders of the smaller numbers of shares were Ernie Wright, C.T. Wilson, William Ronald Cargill, Henry Ariano, John Nagle, John Easter Leece and Reginald Cook Dickens.

The Patricia Transportation Company tore down the Triangle Fish office that they were using and built a new office and warehouse with additional docks. Next to the east of the warehouse a stiffleg derrick was erected for the unloading of heavy equipment, etc. from the railway cars directly into scows. The company had three loading gangs: one operated the derrick for unloading heavy equipment directly into scows; another was a chute from the railway cars to load directly into scows merchandise such as dynamite, beer, flour, sugar, etc.; and, the third gang handled the unloading of miscellaneous wayfreight from railway boxcars onto a truck which was then hauled to the dock for unloading into scows.

During this period, both Patricia Transportation Company and Starratt Airways and Transportation Company installed diesel-powered generator light plants to produce 25 CD electrical power for lighting up the area of their waterfront facilities. Lines were also run to the respective employees' homes who requested this service. However, this private electrical service did not last long as Ontario Hydro came through with their 60 cycle service in 1939.

In the spring of 1937, the company was purchased by James Richardson & Sons, Limited of Winnipeg. This led to to another period of expansion for the company.

In 1941, R.W. Starratt sold his company to the Canadian Pacific Railway Company and since they were only interested in his aviation assets, Patricia Transportation Company acquired Starratt's surface assets in January 1942. During 1942, the CPR was buying out all the various aircraft companies throughout Canada for the establishment of their Canadian Pacific Airways' subsidiary.

The Patricia Transportation Company moved into Starratt's office as it required additional space for its office personnel due to the increase in tonnage volume. When Patricia moved out of their office and warehouse facilities, Cochrane Dunlop Hardware Company moved into those two buildings.

In 1951, the company transferred its head office from Hudson to Winnipeg. Foreman Allan Sandin was left in charge of navigation until the fall of 1953, when hauling to Doghole Bay ceased.

The company's operations continued to wind down with the construction of Highway 105 to Red Lake from Vermilion Bay, and Highway 599 from Savant Lake to Pickle Lake area. Winter freighting contracts also declined in this period. The Patricia Transportation Company was finally phased out in 1964, after 33 years of operations.

Opposite Page: The wheelhouse of the* Comet *at Red Lake with Mel Parker at left and Jack Wish.

APPENDICES

GLOSSARY

Caboose — the final unit of a tractor swing, which housed the cooking and sleeping facilities.

Claim — a tract of land staked out for future development, such as a mining claim.

Glass-out — occurred when the water on a lake was so calm that it looked like a mirror, making it difficult for a pilot to land an aircraft due to the lack of depth perception. The same problem could arise in the winter if there a smooth coat of unbroken snow which made it difficult for a pilot to judge altitude when coming in for a landing.

Jack-knifing — to turn and form a 90 degree angle, for example, a tractor towing a trailer which slid together due to icy road conditions.

Muskeg — a spongy, unstable bog over which travel is difficult and upon which building is problematic.

Permafrost — a permanent frozen layer of ground, common in the far northerly reaches of Canada.

Portage — the carrying of boats or other goods from one body of water to another or around an obstacle, such as rapids.

Scow — flat-bottomed, open-air boat capable of carry large quantities of freight. Was generally towed by a tugboat or other motorized craft.

Swing boss — the person in charge of a winter tractor freighting crew.

Tractor swing — The main means of transporting heavy goods to remote customers during the winter. They consisted of one and usually two tractors, each towing several sleigh loads of supplies and fuel over frozen terrain. In the event one tractor went through the ice or slid off the trail, the other tractor could be used to winch it out.

Windfall — a blockage on a pathway, such as dead trees blown down by the wind.

APPENDIX I
Boats Utilized by the Patricia Transportation Company

The following is a list of boats operated by Patricia Transportation Company of Hudson, Ontario before the acquisition on January 1, 1942 of additional boats from the Canadian Pacific Railway Company when they purchased the Starratt Airways and Transportation Company Ltd., also of Hudson.

Archibald - Length 42 feet, beam 12 feet, powered by a three cylinder Calenberg semi diesel engine which was later replaced by a 360 GMC two cylinder diesel engine.

Arrow - A motor launch.

Canvul, Comet, Patricia, Standard - There were four identical boats built by Ole Gustafson and launched in 1933 for operation in the Chukuni River waterway system. Their lengths were between 32 and 35 feet, and the beams were 9.5 feet. *Canvul* and *Comet* were powered by 40 Hill diesel engines. *Patricia* was powered by a 40 HP gas engine. *Standard* was powered by a six cylinder, 60 HP Hill diesel engine.

Flatty #1 - Length 30 feet, beam 8 feet, powered by 40 HP Buffalo engine.

Hexagon - Built by Olaf Gustafson and launched in 1934. Length 42 feet, beam 12 feet, powered by a four cylinder Calenberg semi diesel engine.

Lac Joe - Built by Ole Gustafson and launched in 1936, which was one of the last boats built by him. Length 45 feet, beam 9.5 feet, powered by a 110 HP Gray marine diesel engine.

Marie - A motor launch.

Octagon¸ Pentagon - These two steel boats were purchased from Russel Bros. of Owen Sound, Ontario in July 1946. Length 48 feet, beams 12 feet and both powered by 360 GMC two cylinder diesel engines.

Triangle - Built by Ole Gustafson and launched in 1920. Lengthy 48 feet, beam 12 feet and powered by a three cylinder Calenberg semi diesel engine.

Wapesi - An Indian name for swan. Built by Ole Gustafson and launched in the spring of 1935. Length 65 feet, beam 18 feet with two decks and hold below. The first deck carried approximately 12 tons of freight. It was powered by a 205 HP Cummins diesel engine. It was equipped with two life boats to carry 12 persons each and was licensed to carry 24 passengers and crew.

APPENDIX II
Crawler Tractors

The following is a list of the type of crawler tractors used by the Patricia Transportation Company during their winter freight hauling operations in northern Canada.

Cleveland Tractor Co. — Tractors

In the late 1920's and early 1930's, Cletrac 25's, 30's, 35's and 55's gas tractors were used. The Cletrac 55 was a powerful machine but, by the same token, it was a gas guzzler. About 1934, the company came out with their diesel tractors and Patricia purchased several of their Cletrac 40's.

Cletrac 40 - 6 cylinder, 40 HP diesel engine; weight 10,000 pounds; speed about four miles per hour.

Caterpillar Tractor Company — Tractors

D6's - 6 cylinder, 55 HP diesel engine; weight 16,000 pounds; speed about four miles per hour.

D7's - 4 cylinder, 81 HP diesel engine; weight 25,340 pounds; speed about five miles per hour.

D8's - 6 cylinder, 130 HP diesel engine, weight 36,310 pounds; speed about five miles per hour.

International Harvester — Tractors

TD40 - 4 cylinder, 40 HP diesel engine; weight 24,000 pounds; speed about four miles per hour.

TD14 - 4 cylinder, 90 HP diesel engine; weight 16,000 pounds; speed about five miles per hour.

TD18 - 6 cylinder, 140 HP diesel engine; weight 24,000 pounds; speed about five miles per hour.

Jim Carson, master mechanic, stated that Caterpillar Tractor's D6's were ideal machines for most jobs as they were heavy enough to pull good loads and light enough to carry them on six inches of ice, while the D7's were the workhorses as they were four tons heavier.

APPENDIX III
Tractor Sleighs

Lombard sleighs were used on the Berens River and Lynn Lake and other freighting projects. They were far better engineered for winter freighting operations than the McLaren sleighs that were used in the Hudson area operations during the early days.

The cross arm system on the Lombard sleighs was far superior to the McLaren sleighs with their tongue (pole) and cross arm chain method. If a cross arm broke, it could be readily replaced with a new one, as all the brakeman had to do was pull the pin that anchored it to the cross arm, slip a new one in and put the pin back in again.

If slush conditions were encountered and it was necessary to uncouple the sleighs, all that the brakeman had to do on a Lombard sleigh was pull the pins anchoring the cross arms. When a similar situation developed with McLaren sleighs the brakeman had to lie down with his elbows up to his shoulders in slush to uncouple the tongue (pole) of the sleigh fastened to a clevis at the centre of the back bench of the sleigh.

If it was necessary to lengthen the sleigh for a longer rack for hauling of items such as pipe, steel drilling rods or lumber, the brakeman just changed the length of the cross arms on the Lombard sleigh. The McLaren sleighs used chains for the cross arm system so that it was necessary to let out so many links to lengthen the sleigh. This was awkward for a brakeman to get at in order to release the required number of links.

The best feature of the Lombard sleighs was the fact that with their cross arm system, the track of the first sleigh runner would be followed in the same track by the runner on the last sleigh regardless of the number of sleighs pulled. Six feet was the standard length of the cross arms on the Lombard sleighs. However, seven, eight or nine foot cross arms were used depending on the length of material to be transported.

In the case of McLaren sleighs, there was a certain amount of play in each tongue (pole) area causing each sleigh to swing out and cut a new track, especially on a curve. The swing out of the sleigh in such cases might be as much as a foot causing an extra drag on the pull for the tractor.

APPENDIX IV
Gold Mines

Red Lake Area

John E. Richthammer's The End of the Road *gives detailed information on producing mines in the Red Lake area. As well, D. F. Parrott's* The Second Gold Rush to Red Lake 1946 *provides comprehensive coverage of all aspects of gold mining in the area, especially the Campbell Lake Gold Mine of Balmertown, Ontario. And Rae Kiebuziski's* Yesterday the River — History of Ear Falls District *also examines the prospecting and gold rush activities in the area.*

Howey Gold Mines

This was the first producing mine in the area. The Patricia Transportation Company had an exclusive freighting contract from the beginning of the 1933 navigation season until the mine was abandoned in 1941.

Brothers Lorne and Ray Howey, along with partners George McNeeley and W.P. Morgan, arrived in Hudson from the east in May, 1925 with the intention of prospecting in the Red Lake area to satisfy their curiosity of the hidden mineral riches.

On July 25, 1925, Lorne Howey stayed in camp due to a slight axe injury to his foot. His partner, McNeeley, was poking around the rocks east of their campsite when he stumbled upon a quartz stringer where a large tree was blown over, its roots torn from the rock it had been growing over. He ran back to Lorne with a sample, who panned a trace of gold from it. They went back to the site together to examine it more closely. Lorne then followed it to the south where it widened considerably and showed promise in visible gold.

At the same time, Ray Howey and Morgan were prospecting nearby. Ray then made a find, which was on the same vein. The excitement of the two brothers finding gold at the same time made them literally leap with joy.

It turned out they were working along a vein that was eight feet wide in some places and 41 feet in other places. Lorne and McNeeley staked 13 claims while the others staked out nine more for the McIntyre Interests as Ray and his partner were sponsored by them, while Lorne and his partner were on their own.

In early September of 1925, the group returned to Hudson and then to Haileybury to record the claims and seek the help of a mining prospector. After the assays, Lorne Howey made a long distance telephone call to mine promoter Jack Hammell in Toronto.

Hammell agreed to look at the find first before making any deals. He and his mining friend Alex Gillies made a quick visit to Red Lake and after examining the find he was elated over it.

Hammell formed The Howey Gold Mines Limited Syndicate, splitting its shares among Lorne Howey and his financiers. Hammell spared nothing in forming the Red Lake Syndicate. He purchased half of the units of the Syndicate and provided $50,000 for development work.

Dome Mines took a 75 per cent option late in 1925 on the Howey property and started an intensive program of trenching, sampling and diamond drilling. The results from the 18 drill holes were not encouraging and in August 1926, the company dropped the option. When Dome dropped their option, everyone felt it was the wrong move. Doug Wright, their engineer, was infuriated, quit and joined forces with Jack Hammell. Morale in the Red Lake camp had reached an all time low when Dome gave up the property in August 1926.

In the meantime, Howey Gold Mines was incorporated on March 12, 1926 to acquire the assets of the Syndicate under the direction of J.E. Hammell as president and H.G. Young as manager.

The shaft and mine workers were all in Heyson township on the boundaries between claims staked by Lorne Howey in the summer of 1925 for the Howey Red Lake Syndicate. A large acreage was cleared around the mine area and bunkhouses for 75 men were built. Work started in the spring of 1927 and by October a three compartment shaft to 521 feet was completed. Steam power was used to sink the shaft.

In 1927, brothers George and Colin Campbell of Northern Transportation obtained an 800 ton freight contract from Howey Gold Mine to be hauled from the Canadian National Railway railhead at Hudson to their property in Red Lake. They rented scows in Hudson, loaded them with Howey freight and towed them to Goldpines on the west end of Lac Seul when they were unloaded at Ear Falls to be hauled across the lakes in the Chukuni River waterway system and over the four portages on the route. The freight had to be reloaded at each portage before it finally reached the Howey Gold Mine property.

In the spring of 1927, Hammell was successful in locating financing. He appointed Horace Greely Young as the first mine manager. Young was a mining engineer experienced in mine management at several mines in Canada prior to his Howey appointment.

In 1928, Northern Transportation (Starratt Airways and Transportation Co.) bid successfully for moving Howey freight from Goldpines and lost the contract in 1933 to their competitor, Patricia Transportation Co. Ltd., for freighting from the Hudson railhead to the mine property.

During the summer of 1928, Howey applied to the Ontario government for hydro electric power. Arrangements were made to install a generating plant as soon as the dam was completed at Ear Falls, which occurred late in 1929. The power was fed into the system and into Howey on Christmas morning that year.

Just before the mine was to go into production, the October 1929 stock market crash caused their share value to drop to almost nothing. The mill was virtually complete, with no money to pay for it. They were very fortunate in obtaining a cash loan of more than one half million dollars from William S. Cherry, a wealthy Canadian-born Rhode Island merchant, which saved the mine. Cherry was also one of the mine's directors.

A mill which started off with a capacity of 500 tons per day was increased to 900 tons by 1932 and to 1,250 tons by October 1933. It continued at this rate until November 3, 1941, when the mine was closed. The company had planned to continue operations until the summer of 1942 when it was believed the profitable ore reserves would be exhausted. However, on the morning of November 3, the shaft pillar started to cave and after four days of salvage operations, the mine was abandoned. The caving culminated on the evening of the 8th when severe shocks were felt in the neighbourhood. For 11 years and seven months the company mined a quartz porphyry dike containing small veins and stringers of auriferous quartz. The operation was famous as one of the first successful low grade gold operations in Canada. The company name was changed in 1949 to Consolidated Howey Gold Mines Ltd..

The Howey mine became the first producing mine in the Red Lake area on April 2, 1930 and poured its first gold brick May 14, 1930. The first brick poured was valued at $20.690 per ounce and the last one in 1941 was valued at $38.149 per ounce or an average of $31.045 per ounce in Canadian funds. Five million shares were issued and $1,950,000 in dividends was paid to the shareholders. The average recovery of gold per ton hoisted was 0.0817 ounces and the total gold and silver production was $13,167,144.55, which included $12 million worth of gold.

Hasaga Gold Mines Limited

In the late 1930's when Howey Gold Mine was approaching exhaustion, Jack Hammell purchased the adjoining claims originally staked by Ray Howey for McIntyre Porcupine Mines in 1925. McIntyre had spent $65,000 on surface exploration in 1927 but did not find sufficient ore to warrant underground development.

On May 18, 1937, Hammell purchased the entire group of mine claims for $110,000. Surface work began immediately and by September results were so successful that Hammell ordered two separate shafts be sunk.

The Hasaga mine was incorporated October 18, 1938, and by November 26th, the ore was being hauled one mile from Shaft No. 1 in Patricia Transportation dump trucks to the Red Lake Gold Shore mine for milling. The Patricia trucks operated around the clock hauling approximately 361 tons of ore per 24 hour day. Patricia Transportation hauled 124,533 tons of ore for their fiscal year ended November 1941 and 135,968 tons for the fiscal year ended November 1942, or an average of 130,250 tons per year. The company hauled approximately 1.8 million tons of ore during the 14 years of the Hasaga mine's operations.

The mine produced $8.2 million of gold throughout its 14 year history.

Madsen Red Lake Gold Mine Limited

Marius K. Madsen re-staked Faulkenham's claims November 1934 on behalf of the Falcon Gold Syndicate. Madsen Red Lake was incorporated March 1, 1935 to acquire the holdings of the Falcon group. In 1937 a new shaft was sunk at Madsen and early in 1938 mill construction was underway. The mine was located seven miles southwest of Red Lake near the end of highway 618.

Seven miles of road was constructed by the Patricia Transportation Company whose construction road foreman was Arne Groneng, with Slim Dart taking care of the clerical work. Road construction in 1937 from Red Lake to the mine site was supervised by Mr. Christopherson of the Department of Highways in Kenora. The Patricia Transportation Company also secured the hauling of the mine's freight from the beginning of the mine's operation until it was abandoned.

Mine production started in 1939 and was abandoned June 30, 1976. Madsen mine became one of Red Lake's highest and most consistent producers at a time when one was badly needed. It produced nearly $84 million of gold bullion during its 37 years of operations. The mine was sold to Bulora Corporation in September 1974, but due to mine fatalities and depletion of ore, operations ceased on June 30, 1976.

The McKenzie Red Lake Gold Mines Ltd.

Ole Gustafson, Wilfred Wright, Ole Sand and four other partners filed claims during the 1922 gold rush. They spent much money on their claims after finding visible gold. However, due to funding problems, they were unable to develop the property and the claims were dropped.

In 1925, the claims were re-staked and the new group discovered the main shear zone surface outcropping in 1928. Coniagos Mines took an option on the claims in 1931 but dropped them later.

In 1932, McKenzie Red Lake Mines was formed. Milling began in 1935 at a rate of 125 tons per day, which rose to 230 tons per day in 1936.

To the end of 1962, a total of 2,068,036 tons of ore had been milled, producing $25 million of gold bullion. The mine closed in 1965 after 30 years of operations.

Campbell Red Lake Mines Limited

The property was originally staked during the gold rush of 1926 by Major Cunningham Dunlop but he could not obtain finances and the claims went open after his death.

During the intervening years, the claims were staked and restaked by many others. These included George W. Campbell who restaked the claims in January 1944 when he finally discovered gold after a 20-year search in the area. He sold the claims to the Dome group the same year for $60,000 and other considerations.

The mine, located in Balmertown, Ontario, consists of five claims covering an area of 200 acres. It is west of Red Lake and is accessible by a 10-mile road or six miles by boat from Red Lake.

The mine went into production in June 1949 and mills 1,078 tons of ore per day. From 1949 until December 31, 1986, it milled 10.4 million tons of ore, producing 6.2 million ounces of gold valued at $1.2 billion dollars. During 1986, it produced 229,182 troy ounces of gold bullion valued at $81.3 million dollars at a cost of $148 per ounce, one of the lowest cost lode gold in Canada.

It is the main gold mine in the Dome group of mines and is the richest mine in Canada. The mine had proven probable ore reserves of 7.6 million tons or the equivalent of an 18-year supply. It was the last mine brought into production in the Red Lake area.

The mine spent $12 million in 1987 to computerize the mill grinding circuits to increase production by five per cent, and production is expected to increase to 240,000 ounces annually by 1990. Stewart M. Reid is general manager, overseeing some 454 employees.

The Cochenour-Willans Gold Mines Ltd.

The mine was incorporated on December 2, 1938 and poured its first gold brick in February 1939. The mine closed permanently in 1974 after producing $40 million worth of gold bullion during its 35 years of operation.

Dickenson Mines Limited

The mine started production in January 1948 with an initial output of 150 tons per day. It is still in operation and, as of December 31, 1972, it produced $64.5 million worth of gold.

Gold Eagle Gold Mines Limited

Gold Eagle Mine started gold production October 11, 1937 and closed September 14, 1941 after producing $1.5 million worth of gold bullion during its four years of operation. The name was changed to Goldray Mines Ltd. in 1959.

McMarmac Red Lake Gold Mines Limited

The mine went into production October 18, 1940 until sometime in 1944 when it was forced to close because of labour shortages during World War Two. It was reopened after the war in 1947, but closed once more in late 1948.

Red Lake Gold Shore Mines Limited

The mine was incorporated in December 1927 and in 1928 surface explorations were carried on. A steam powered mining plant was installed in 1929. The mine was refinanced in 1934, a 550-foot shaft sunk and a small mill constructed in 1935. The shaft was deepened past 700 feet, but the mine was abandoned August 28, 1938 due to a lack of ore. During its two years of production it produced $750,000 worth of gold. The plant was bought by Jack Hammell in October 1938 and was used for milling Hasaga Gold Mine ore.

Starratt-Olson Gold Mines Limited

The claims were staked in 1926 adjacent to Madsen Red Lake Gold Mine property but were dropped. The claims were restaked in 1934, but dropped until Hasaga took over.

In 1945 Starratt-Olson Gold Mines was incorporated to develop the property. Uchi Gold Mine's Mill was brought in and production began September 1, 1948. The mine operated until 1956 and during its years of operation produced $6 million worth of gold.

The mill was purchased by H.G. Young and operated on its site from 1960 to 1963.

H.G. Young Mines Limited

The mine was incorporated in 1946 in order to obtain claims that were under the water of Balmer Lake.

Exploration work was carried on until 1959 when the company purchased New Campbell Island Mines' claims. In co-operation with Campbell Red Lake Gold Mines, a shaft was sunk on Campbell's property to mine the ore from under the lake.

On September 1, 1960, the first load of ore was trucked the 16 miles to the Starratt-Olson mill for processing. This method of mining was carried on until March 1963 when the mine closed. The mine produced $5 million worth of gold. In 1980, Campbell Red Lake Mines purchased the property from H.G. Young Mines Limited.

Woman Lake Area

Many mines in the Woman Lake area collapsed financially owing to the high cost of operating in so remote an area. Some properties held insufficient ore. One small vein and scattered pockets of gold would not support the initial outlay. Other properties were rich but lacked funds. Managers and promoters alike faced the heartbreak of seeing a dream of a lifetime fade away in spite of the rich grade of ore bodies.

Several mines in the Woman Lake area came into production but had very short lives, lasting from a few months to a few years due to excessive freight costs and other operating expenses for which the value of ore was insufficient for a profitable operation. In contrast, some 15 gold mines in the Red Lake area operated for many years, some as long as 35.

Bathurst Gold Mine

The Bathurst mine, five miles north of the Duncan Mine, experienced the same catastrophe as the Duncan Mine (see below).

The mine was taken over by Sol D'Or Gold Mines Limited and operation resumed. During 1932, 400 tons of ore worth $10,000 were milled from a shaft 40 feet deep and 250 feet long. However, after one year's operation, the mine was closed down again.

Sol D'Or Gold Mine reopened the mine in 1935 and planned to sink a deeper shaft. Although better gold values resulted after sinking their shaft 160 feet, it closed at the end of 1935.

Casey Summit Gold Mine

The mine, located about 70 miles northeast of Goldpines and 100 air miles from Hudson, opened in 1929 and closed in 1930. In 1932, it reopened and the shaft was extended to a depth of 200 feet.

The mine was bought by Argosy Gold Mines Limited and a second shaft was sunk 340 feet from which in 1936 some 9,800 tons was milled, valued at approximately $130,000.

Argosy Gold Mines Limited's name was changed to Jason Mines Limited in 1938. Although under new management the main shaft had been sunk to 545 feet and No. 2 shaft to 339 feet, the mine closed in 1943.

During the four years it operated, 146,068 tons of ore were milled with a value of $2.3 million. The mine reopened in November 1945. The main shaft was sunk to 871 feet and during the summer of 1946 went into full production. It operated until 1951.

Corless Patricia Gold Mine

The property is located about 40 miles northeast of Goldpines. They bought the West Red Lake Gold Mine plant, which was dismantled and delivered to their property by the Patricia Transportation Company during the winter of 1935-36. Due to a lack of funds, the plant was not assembled after delivery and the property was abandoned.

Duncan Gold Mine

In 1929, Duncan Mine on Narrow Lake showed free gold which could be seen on the surface, and after sinking a 100-foot shaft, they missed the vein. Horizontal tunnels were dug off the main shaft but the vein of gold could not be found. They ran out of funds for equipment and on March 22, 1930, the mine closed.

Jackson Manion Gold Mine

In 1932, J.M. Consolidated Mines reopened the property but it was closed permanently in 1936. The manager of the mine, unsuccessful with managing operating funds, became so depressed that he committed suicide. The mine was finally dismantled by Vic Nymark and partners and delivered to new owners.

Uchi Gold Mines Limited

The mine is located about 50 miles northeast of Goldpines, and was started in 1936, financed by Jack Hammell interests.

The first swing of freight to the Uchi Mine was delivered in 1937 by Starratt Transportation Company. They had the freighting contract from the beginning of mine operations until the end of 1941 when their assets and the mine's freight hauling contract was taken over by Patricia Transportation Company on January 1, 1942.

During 1942, Patricia Transportation hauled a total of 48,145 tons of ore from the mine's No. 4 and 5 shafts to the mill at 21 cents per ton. The No. 4 shaft was one mile from the mill and No. 5 was two and one half miles.

Also that year, Patricia transported Uchi miners from the bunkhouses to the shafts and back. Some 23,730 tickets were sold at a price of 12 and a half cents per ticket.

During the winter of 1942 Patricia hauled 460 tons of freight by trucks from Hudson to the mine as well as 31 tons by tractor. During the summer of 1942, a total of 1,551 tons were hauled to the mine by boats and trucks.

Due to the shortage of labour and supplies during World War II, the mine was forced to close at the end of 1942 and the area became a ghost town. The mine produced $4.4 million worth of gold during its six years of operations.

In 1936, the plant machinery and mill was sold to Starratt Olson Gold Mines in Red Lake, another of Jack Hammell's interests. The Starratt brothers, Dean and Don, of Hudson obtained the contract for dismantling the mill and arranging for its delivery to the Starratt Olson property in Red Lake.

Pickle Lake Area

Although two very good gold discoveries were made in the late 1920's, the prospecting activity declined in later years. It is believed this was probably due to the high transportation costs, as all perishable supplies such as meats, dairy products, fresh vegetables and so forth had to be flown in throughout the year from the railhead to Pickle Lake Landing. It was not until the 1949 navigation season that a meat run was established from Hudson to Doghole Bay at the eastern end of Lake St. Joseph and then by truck to the mines. However, this service only lasted until navigation ceased at the end of 1953, when Highway 599 from Savant Lake railhead to Pickle Lake was built.

Central Patricia Gold Mines Limited

The mine was located in Connell Township about 20 miles northeast of Lake St. Joseph and 90 miles north of Savant Lake, Ontario on the Canadian National Railway's main line.

During 1929 and 1930, considerable exploration and development work was carried out. The mine was organized in 1931 and started production in 1933.

Between 1932 and 1951, surface and underground diamond drilling of about 203,000 feet and 217,000 feet respectively were carried out and a shaft sunk to a total depth of 2,226 feet at level intervals of 125 feet to the 1,000-foot level, then 150 feet down to 2,050 feet vertical winze; No. 3 from 2,050 to 3,722 feet and, vertical winze No. 4 from 3,400 to 4,020 feet at level intervals of 150 feet in both winzes. A total of 11,266 feet of crosscutting and 50,860 feet of drifting was also carried out during the years of the mine's operations.

Its No. 2 mine operations were located four miles east of the main shaft. The shaft at this mine was sunk to a depth of 1,024 feet with seven levels, mostly at intervals of 150 feet, drifting 3,741 feet and crosscutting 1,775 feet, and 16 underground drill holes totalled 1,970 feet.

From 1938 to 1940, the No. 2 mine milled 18,886 tons of ore and recovered 13,158 ounces of gold valued at $477,965. When the operations of the main mine, including No. 2 mine, were closed in 1951 due to a lack of orebodies, a total of 1.8 million tons of ore was milled, producing 621,806 ounces of gold and 58,349 ounces of silver for a total value of $23 million. Its shareholders received a total of $4.7 million dollars in dividends.

The mine amalgamated with McVittie-Graham Mining Co. Ltd. on June 12, 1980 to form Central Patricia Limited and then Central Patricia Limited amalgamated with Con-West Exploration Co. Ltd.

Pickle Crow Gold Mines

The mine was discovered in 1928 about 161 miles east of Red Lake, in the Patricia Mining Division, by the prospectors flown out to the field by the Northern Aerial Minerals Exploration Co. Ltd. (NAME). The find was one of their most important discoveries.

In March 1933, the mining plant was hauled by Starratt Transportation Company tractors from Savant Lake, Ontario to the mine site. The mine went into production in June 1935 and 10 months later it started paying dividends to the shareholders. In 1938, the Pickle Crow Gold Mines took over the Albany production and tripled the capacity of its mills to 400 tons a day.

Jack Hammell organized the company on his own financing efforts. He was the largest shareholder and he is reported to have received $8 million in dividends from the mine before his death in 1958. The mine closed in 1966 and during its 30 years of operation, it milled three million tons of ore which produced 1,446,214 ounces of gold and 168,757 ounces of silver valued at $35 million.

Pickle Crow Gold Mines Ltd. sold its assets to Highland Crow Explorations Ltd. in 1968 who in turn were succeeded by Highland Crow Resources in 1977. Pickle Crow property consisted of 98 contiguous patented mining claims covering 3,912 acres. The property is owned by Noramco Mining Corporation which has 100 per cent working interest in the property.

From 1981 to 1984 Gallant Gold Mines Limited, under option from Highland Crow, carried out geological and geophysical surveys on the property and other properties in the area. They ended with the drilling of 25,052 feet of surface diamond drilling. In 1985 Highland Crow terminated the option and spent $1 million recompiling the data from the mine and resampling the Gallant core. Results were promising and by March 1986, 21,250 feet of diamond drilling was completed in 50 holes.

In December 1987, Noramco Mining obtained the property through the amalgamation with Highland Crow Resources and Emerald Lake Resources. The proven reserves are approximately 300,000 tons, probable reserves amount to two million tons and possible reserves of nearly five million tons, for a total of 7.3 million tons.

APPENDIX V—TOWNS

HUDSON

Hudson is located on the south shore of Lost Lake in the Vermilion Township Addition, in the Patricia Portion of the District of Kenora in the northwestern part of Ontario. It is 252 miles east of Winnipeg and 14 miles west of Sioux Lookout, on the Canadian National Railway transcontinental main line. It is now accessible by the 10.5 mile all weather Highway 664 connecting with Highway 72 to either Sioux Lookout for another five miles or Dryden via Highways 72 and 17 of approximately 65 miles. It is surrounded by forest with the exception of the north side, which is the shore of Lost Lake.

Pioneering Days

When the railway from Montreal to Winnipeg was completed in 1910, Hudson's Bay Company personnel were the first to take advantage of its location as they immediately established a trading post to deal with the Indians in trading furs for supplies. They also moved their railhead distribution facilities to Hudson from Dinorwic on the Canadian Pacific Railway, which was completed in 1896, as it was much closer to their Lac Seul Post in the outlying areas further north.

Hudson was in complete isolation, and the only way in or out was by train. There was a tri-weekly freight train service between Winnipeg and Sioux Lookout. The Canadian National Railway also had a weekly passenger train service. It was not until the Red Lake gold rush of 1926 that good passenger service was established.

Mrs. Maria Lukinuk related that when she arrived from North Pines on September 4, 1919 with her husband, who was a section foreman for the railway, and their six month old baby, she was the only white woman in the area. They lived in the two storey section house provided by the railroad.

Water had to be carried from Lost Lake for cooking, washing, etc. and in the winter, snow had to be melted. It takes a lot of snow to make a barrel of water. In later years, water was hauled by a team of horses to various homes for 25 cents per barrel or individuals hauled it by toboggan. It was not until 1960, when private individuals started on a running water system, which was finally taken over by the Ontario government 27 years later.

Bucksawing wood was a constant chore during the winter. In order to heat the Section House building, there was a big round stove in the front room in which a mixture of green and dry wood was used. Dry wood was used in the cookstove and it took at least a half hour to get it hot enough to cook a breakfast of porridge, bacon, eggs and coffee.

The Hudson's Bay Company post did not carry a variety of groceries, etc. as their main business was with the Indians. So, certain quantities of groceries had to be ordered from Winnipeg and sometimes they went to Sioux Lookout by railway motor car to shop. It was not until the middle 1920's that several grocery businesses were established.

Aircraft came into use on March 3, 1926 and an all weather highway to Sioux Lookout was completed in 1937. The only communication to the outside world was by Canadian National Telegraph. A private telephone company ran a line from Sioux Lookout in 1937 and finally the Bell Telephone Company extended their service to Hudson in 1940.

From 1926 to 1934 there was considerable activity on the waterfront. However, when the price of gold jumped to $35 per troy ounce in 1935, the boom in freighting to Red Lake accelerated. It was also the year that the Central Patricia Gold Mines and Pickle Crow Gold Mines started their operations, approximately 25 miles northeast of the end of Lake St. Joseph. The result was that tremendous building activity took place on the waterfront to handle the increased volume of tonnage.

Entertainment and Recreation

During the winter of 1935, a Hudson Dramatic Society Group was formed and the first three act drama was performed under the direction of Major Chetwyn. Performing in *For Whom the Bell Tolls* were Don Patterson as the father, Mr. Sutherland as the beggar, Jack Wish as the mandarin, and Mmes. Coote, Duklow and Wish as the Tingley Sisters.

In the winter of 1936, another three act comedy drama was performed under the direction of James A. Tickner, entitled *Deacon Dubbs*. Its cast included Baden Webb as Deacon Dubbs, Mrs. McLennan as Miss Philipena Popover, J. Morrison as Amos Coleman, Mrs. Shaver as Rose Raleigh, F. Fahlgren as Rawdon Crawley, Miss Scouten as Emily Dale, Jack Wish as Major Moses McNutt, Mrs. Wish as Trixie Coleman, Jack Atkinson as Deuteronomy Jones and Mrs. Bea Pinkess as Yennie Yensen.

These two plays were presented at Sioux Lookout and Dryden after first being performed in Hudson.

Dances were held every Saturday night at the community hall. However, these were held at different times than normal. Instead of starting the dance at 8 p.m. and ending around midnight, the Hudson dances started around midnight and lasted until daybreak on Sunday morning.

Colonel J.M. Candlish of the Canadian Legion arranged to procure movie reels which were shown every week in the community hall.

Summer activities consisted of baseball, tennis, boating and fishing, while during the winter months there was hockey, skating, tobogganing and moccasin hiking parties. Badminton was also played in the Community Hall and curling took place on the lake.

In 1937, some movie celebrities from Hollywood arranged to use Joe Keneally's Lodge facilities for fishing on Vermilion Lake for a week. They certainly came prepared to rough it by renting a Canadian National Railway's Pullman car with a porter and a dining car with a chef for their use while at Hudson. These two railway cars were spotted on the railway siding at their disposal. Some stayed at Keneally's Lodge for the night while others came back every evening to their railway quarters. Among this group was Andy Devine, who made quite a hit with the Hudson residents. It is presumed that the reason he got along so well with the local people was the fact that he came from the small town of Kingman, Arizona, after whom the town named a street called "Andy Devine Boulevard" in honour of its native son.

Perspective

The population of Hudson was about 600 people at the height of the transportation trade and the population of the town is still about the same today, made up of approximately 200 McKenzie Forest Products employees and their families as well as descendants of pioneers and newcomers.

Hudson, situated on the south shore of Lost Lake had the unique position of being the only navigation site located on the main railhead line for the outlet to the north. Hence, it made a major contribution in the development of the mines and towns in the Red Lake and Pickle Lake areas. If the excellent navigation access to these two areas were not available from Hudson, it is possible that the production of the various gold mines would have been delayed for years or they would not have been developed at all. For example, the Howey Gold Mines, the first producing mine in the Red Lake area, produced gold from a very low grade of orebodies. The Red Lake area mines could have met the same fate as those in the Woman Lake area that were unable to carry out the development of the properties due to the very high cost of operations, particularly transportation.

Throughout the years, Hudson enjoyed a booming economy as a result of being the base for commercial fishing, transportation — both surface and air — as well as forestry and tourism. Commercial fishing was phased out in 1979 after 60 years of operations, while transportation was phased out in 1953 after 28 years of operations. Although forestry operations were not carried out on as large a scale as in the past, the McKenzie Forest Products facility is still operating in the area, employing 200. Tourism will continue to grow due to its excellent location on Lac Seul, Lost and Big Vermilion Lakes, whose waters are teeming with all kinds of fish.

Goldpines

Goldpines is located on the northwest shore of Lac Seul approximately four miles south of the Ear Falls Dam. In 1912, the Hudson's Bay Company built a fur trading post at this location which was called Pine Ridge at the time, but later renamed Goldpines in 1926 as the Canadian Post Office already had a Pine Ridge post office in their system. Howard Halverson was its post manager in 1923.

With the Red Lake Gold Rush of 1926, the town's economy started to boom as it was the end of the first 110 mile navigation lap of the route to Red Lake which was covered in one day. Passengers had to disembark for Red Lake and stay in the hotels overnight before proceeding on to Red Lake. Freight had to be unloaded and portaged to the English River side for furtherance to Red Lake. Hotels, stores and other facilities were built to take care of the passengers and the handling of freight.

The construction of the Ear Falls Control Dam by the Lake of the Woods Control Board in 1928 and the construction of the first unit of Ontario Hydro's electric generating station in 1929 further increased the activity of the community. Another factor that added additional activity to the community was the beginning of the exploration and development of several gold mines in the Woman Lake area, approximately 50 miles further north.

The construction of the dam raised the water level of Lac Seul and flooded out the Hudson's Bay store, so the company built another one on higher ground, but still in the same area, with David Learmonth arriving in December 1929 to be its new post manager. He was transferred in 1934 to Hudson to take over the management of the store in that community and Nichols took over his place at Goldpines.

On July 1, 1929, the town held a sports day. This was the peak year of the town's activity when it had four stores, two hotels, two banks (Imperial Bank and Bank of Commerce), school, laundry, barber shop and a population of 1,000 people.

The town's decline began when the freight for Red Lake was bypassing Goldpines as the Northern Transportation Company (Starratt's) took it across the portage to their base at Little Canada for furtherance to Red Lake. Further decline was experienced when most of the mines in the Woman Lake area closed down, not because of the lack of potential gold ore reserves, but due to the high cost of operations required to finance them, etc.. Heavy mining equipment had to be hauled by tractor trains to the various properties during the three month winter freighting period while all other materials and supplies had to be flown in. The only operating mine left was the Argosy Gold Mines at Casummit Lake, 70 miles north of Goldpines. It was later renamed Jason

Mines, which carried out production on and off until it finally closed in 1953. For example, to illustrate the cost problem, Ontario Hydro did not service the area, so electric power was generated by diesel engines requiring considerable quantities of fuel oil. During the summer of 1935, Canadian Airways had a 600 ton contract to fly the fuel oil in their flying box car from Goldpines to the mine property and during the summer of 1936, they flew in 700 tons of freight to the mine. Therefore, one can just imagine the cost of flying such large tonnages.

The final death blow to the Goldpines economy came in 1931 when the passenger plane fare from Hudson to Red Lake dropped to $25 and air freight was seven cents per pound for every 100 miles. This was of tremendous help to the travelling public as it was almost as cheap to fly to Red Lake from Hudson as was the cost of a boat fare, overnight lodgings at hotels and meals at Goldpines. It also eliminated the day long travel by boats, without any facilities, in the Chukuni River Waterway as well as crossing the four portages on foot. It took two days by boat to make the trip from Hudson to Red Lake via Goldpines whereas it only took one hour by plane.

Goldpines became a ghost town with row after row of empty buildings as the banks, government establishments and various businesses all moved to Red Lake. The float airbase was approximately one mile from Goldpines and although large tonnages of freight were flown to Argosy Gold Mines, it did not help the town's economy very much. It improved slightly from 1937 to 1941 when the Uchi Gold Mines, approximately 50 miles to the north, carried out exploration and development work. However, production closed down in 1942 due to the wartime labour shortage, the high cost of operations and the lack of good ore bodies.

The Goldpines post office closed on October 14, 1950 and the Hudson's Bay Company sold their store to Weir's Recreational Centre which now operates tourist facilities known as Gold Pines Camp. The area has abundant fish, such as pickerel (walleye), trout and many others. The quiet, friendly and relaxing atmosphere of the camp provides one with a well deserved holiday. It is only five miles away by an all weather road from the town of Ear Falls.

Red Lake

Red Lake is situated in the Judicial District of Kenora, in the Patricia portion of northwestern Ontario in the township of Heyson.

Red Lake came into prominence with the discovery of gold by the Howey brothers in the fall of 1925 and, as a result, the Red Lake gold rush began in 1926.

The first gold producing mine was Howey Gold Mines Limited, which started production in 1930 and continued to operate for 11 years when it closed in 1941 due to a lack of orebodies. Other mines were discovered in the area and brought into production and some of them continued producing well into the 1970's.

In 1934, the Patricia Transportation Company built an office, warehouse and docks to carry out their freighting operations to Red Lake and area. These operations carried on for the next 14 years until navigation ceased due to the construction of Highway 105 from Vermilion Bay to Red Lake. The office buildings and warehouse were sold in 1959.

The town of Red Lake had the most disastrous fire on record with the loss of many lives and injuries when the three storey Red Lake Hotel burned to the ground on July 1, 1945. The fire started at 1:30 a.m.. Many guests who jumped out the windows suffered severe injuries, including Ernie Wright, general superintendent and Don McLennan, treasurer, both of the Patricia Transportation Company, who happened to be there on business at the time. The seriously injured were flown out to Winnipeg during the night and the following morning.

Andrew Szura, 34, who worked as a dock hand in the summer and as a tractor driver during the winter for the Patricia Transportation Company, became a hero by saving three people. Seeing the building burning, he dashed into it and dragged one person out to safety. He went in again and dragged another person out and, to the horror of spectators, he again dashed into the inferno and dragged a third person out. After coming out the final time, his hair singed off and his whole body charred, he dropped unconscious on the street and was given first aid on the spot. In the very early morning he was flown to a Winnipeg hospital where he died from the severe burns.

There was no hospital at Red Lake at the time, so some of the injured and badly burned were rushed seven miles to the Madsen Red Lake Gold Mines hospital. Dr. Joe McCammon was the resident doctor who worked during the night and day and was credited with saving many lives. He later recalled, "There were some people so badly burnt that they were almost unrecognizable and there was nothing I could do for them."

In August 1947, the 108 mile Highway 105 from Vermilion Bay to Red Lake was officially opened to traffic. Heavy traffic could move over it in 1948 and, as a result, the navigation operations ceased to Red Lake.

Construction of this highway eliminated Red Lake from complete isolation to the outside world as had been the case in the past during freeze-up and break-up when no air or surface transportation could operate during these two periods. The only exception had been outside communication by radio telephone. While it was a blow to the navigation transportation companies in that their operations ceased, it was a boon to the community as it was no longer isolated and further development of the area's mining industry, tourism facilities and other industries could proceed.

The population of Red Lake in 1985 was 2,189 people with a trading area population of 10,800, which included the communities of Madsen, Starratt-Olsen, McKenzie Island, Cochenour, Balmertown and others. Red Lake continues to be the centre of things, more than just "the end of the road" in Ontario's north. The Ontario government offices, the district hospital, medical clinic and high school, as well as banks and churches, are based here.

Large stands of virgin timber stretch to the north, east and west of Red Lake. They provide the basis for two of the town's economic mainstays, i.e. tourism and timber harvesting. Two large pulp and paper companies and 18 independent timber operators are active in the area. More than 70 tourist camps and lodges are available, ranging from deluxe to no frills cabins. There is an abundance of fish and wildlife. Pat Sayeau, president of the Chamber of Commerce, said that during 1985 these businesses channelled 5,000 guests through a town of 2,189 people.

With its proximity to a major airport at Winnipeg and its large float plane base, the Red Lake tourist facilities are extremely accessible. Red Lake also utilizes the fully equipped airport with a 4,000 foot runway situated on the outskirts of Cochenour to handle all land-based air traffic in the district.

Although there has been no producing gold mine in the area since the early 1970's, exploration activity throughout the mining area is very attractive. In September of 1939, the Berens River Gold Mines at Favourable Lake, some 150 miles northwest of Red Lake, started gold production and continued until September 1949 when it ran out of ore reserves. Hence, there is no reason not to believe that more gold mines or base metal properties will be found in this vast region.

The town has a well-rounded shopping centre and many recreational facilities and with its abundance of electric power, water resources, as well as the tourism and forestry industries, the town will continue to grow and prosper.

Balmertown

Although Balmertown was located approximately 10 and a half miles away from Red Lake in Balmer Township, for all practical purposes it was referred to as part of the Red Lake district. It was later known as the Improvement District of Balmertown, which included the towns of Balmertown, Cochenour and McKenzie Island. The IDB board, with the agreement of neighbouring municipalities, extended the boundaries to twice its size and, as a result, its status was changed to the township of Golden on December 1, 1985. During an election in November that year, a reeve and councillors were elected to govern the new townships.

Its population as of 1985 was 2,200 people and the trading area population was 10,900. Highway 125 links Cochenour and Balmertown to Red Lake. It is the home of two gold producing mines viz. Campbell Red Lake Mines Limited and Arthur W. White. Campbell Red Lake Mines Limited is the richest gold mine in Canada with adequate ore reserves to carry on with production for a great number of years. Its operations are carried out under the capable manager Stewart Reid, whose mine employs 445 people.

Reid said that with the installation of improved efficiency over the last few years that in 1984 they achieved the highest recovery rate ever at Campbell, with 94.5%. He said the improvement was due to new mill equipment and the installation in 1983 of carbon columns to recover gold from tailings effluent.

The other producing mine is Arthur W. White, formerly known as Dickenson Mine, managed by Peter Busse and which employs 244 people.

Exploration and development is being carried out on the old Cochenour Willans property by Esso Minerals. Several miles away within the township, the McFinely Red Lake joint venture is involved in exploration work.

The Red Lake Airport, situated on the outskirts of Cochenour, which handles all land-based air traffic in the district, is equipped with a 4,000 foot runway. Three airlines offer scheduled flights with connections to major centres both east and west. Two other aircraft companies offer scheduled and chartered flights to other northern communities. Charter services are also provided by two helicopter operations with offices at the airport.

The Golden Township has several facilities that serve the entire Red Lake District. One of these is the Owen J. Mathews Manor located in Cochenour which provides accommodation for residents of the area who, through age or infirmity, require some attendant care.

All types of recreational activities are available in the community, including the Red Lake Area Golf and Country Club on the outskirts of town, which has a well kept nine hole golf course.

Pickle Lake

The town of Pickle Lake is located on the northeast shore of Pickle Lake in Ronsford Township in the Patricia District and in northwest Ontario.

The townsite was started in 1928 as Pickle Lake was the closest that planes with floats in the summer and skis in the winter could land to service the mine properties. Central Patricia Gold Mines' property was just one mile away and it was seven miles to Pickle Crow Gold Mines' property. All passengers and perishable freight had to be flown to Pickle Lake Landing from either Hudson or Sioux Lookout for delivery to the two mines throughout the many years of their operations.

This created quite an activity in the townsite with the establishment of the Hudson's Bay store, Davidson's Severn store and Hooker Bros. store, who also had their own aircraft to fly their fish that they bought commercially to the outside. There was a considerable quantity of sturgeon flown out along with other catches.

Mike Debroni operated a bakery as well as a bottling works for supplying soft drinks in partnership with Russell Harasyn. They also operated a taxi service. Mrs. Bresseau ran a coffee shop and there was also a bank and a barber shop.

The Department of Forestry had their station located in the area with J. Guertin in charge of four employees during the summer months. They had a forestry plane on a standby basis at Sioux Lookout headquarters. Upon Guertin's retirement, he moved to Winnipeg.

The Patricia Hotel was built on Central Patricia Gold Mines property by Ernie Wilson, who operated it. He also had the post office located near his hotel. The hotel is still operating under new management and now there is also a motor hotel built in the area. Across the road from the Patricia Hotel there was a Catholic and an Anglican Church and between the two was the jailhouse, with one provincial policeman stationed at Central Patricia Mine.

Pickle Crow Gold Mines also had a nice hotel on their property site as well as a general store. It might be mentioned that Central Patricia Gold Mines also had a general store on their property.

Koval Bros., after moving to Pickle Lake from Savant Lake in 1944 to carry out the trucking from Doghole Bay to the mines, as well as winter tractor freighting from Savant Lake before Highway 599 was completed, built a garage, hotel and restaurant in town. They also operated a sawmill in the area and supplied wood to the mines, as well as timber for their underground work. The Koval Brothers' sons are continuing with the family's trucking business as well as other interests in the area.

Harry Everett, the former Patricia Transportation Company's agent at Doghole Bay, related that there used to be a radio shack between Pickle Lake and Central Patricia Gold Mines' property. It was closed down and was converted to a "chicken house" run by Muskeg Myrtle. The reason they called her "Muskeg" was the fact that her operation was closed down so she moved out to the edge of the townsite and installed a tent where she carried on her business. She had one regular girl and a couple more girls were brought in over the weekends. There were several women raped at the Pickle Crow Gold Mines so Myrtle was brought back to the radio shack. In 1951, she closed her business and got married to one of the miners.

The economy of the town was affected when Central Patricia Gold Mines closed down in 1951 and hit very hard when Pickle Crow Gold Mines closed their operations in 1966. Its economy depends on Koval Bros. trucking operations, mine prospecting in the area and tourism in the summer months.

In 1977, Noramco took over the Pickle Crow Gold Mines property and carried out considerable exploration and development work. Thousands of feet of diamond drilling was carried out over the years, resulting in the discovery of sufficient orebodies to go into production in the fall of 1989. It is not known whether their plans materialized. If they did go into production, it would be a great boost to the town's economy. However, the town's bustling heyday activities when Central Patricia and Pickle Crow were in operation is a thing of the past.

There is a fine beach where the planes land and its current population is approximately 500 people. With all the prospecting activity going on in the area, it is possible that a good gold orebody will be found and a very productive mine will be established and the town will continue to grow.

Savant Lake

Savant Lake is located in the Thunder Bay district in the northwest part of Ontario. It is 312 miles east of Winnipeg on the Canadian National Railway's main line and it is approximately 90 airmiles south of Pickle Lake.

Beginning with the 1930's, John Haverluck and his brother-in-law Jack Koval owned and operated the Savant Hotel and the Savant General Store. Great Lakes Pulp and Paper Company had three large bush camps near Savant which were closed long ago.

The town came into prominence during the winter of 1933-1934 when J.E. Hammell hired Starratt Airways and Transportation Company to haul the mining equipment, supplies and materials to his Pickle Crow Gold Mines' property site located on the Crow River in Connell Township. Starratt also hauled the transformers and other heavy equipment and supplies to the construction site of Ontario Hydro's generating plant at Rat Rapids at the mouth of the Albany River on Lake St. Joseph. It became the railhead for the transportation of machinery, equipment and supplies to the Pickle Lake mining area, some 90 miles away during the winter freighting seasons, for the next 21 years (1933 to 1954). This ended with the construction of all weather Highway 599 from Savant to Pickle Lake, which was completed during the summer of 1954.

During the winter of 1935-1936, Haverluck and Koval started their winter freighting operations from Savant into the Pickle Lake area, using horses the first winter and then diesel tractors hauling sleighs of freight. John Haverluck and Jack Koval sold their general store and hotel at which time Haverluck retired and Jack Koval continued with the winter freighting business.

Subsequently, he retired and turned the business over to his three sons, Alex, Bill and Donny. Donny was in charge of the office and administration of the company while his two brothers took care of the field operations. In 1944, the Koval Brothers moved their headquarters and their families to Pickle Lake.

Joan Zawada, a resident of Savant Lake, related that the former general store owned by Haverluck and Koval burned down in 1986. It was rebuilt by Dennis Mausseau, who has also built The Four Winds motel. Also, since the hotel was sold by Haverluck and Koval, there have been many owners over the years. It is currently owned by Roland Maede of Madison, Wisconsin.

Zawada also related that there is considerable prospecting activity going on in the area north of Savant Lake. There are also three tourist camps on Lake Savant and at least six on Highway 599 between Savant Lake and Pickle Lake. She mentioned that there is very little freight arriving at the railway station since the mines at Pickle Lake have been closed down for years and any freight for the area is handled by trucks.

ARTICLES

Cobb Jr, Charles E., "The Great Lakes Troubled Waters." National Geographic. July 1987.

Newman, Peter C., "Canadian Fur Trade Empire." National Geographic. August 1987.

BOOKS

Bray, Matt and Epp, Ernie, *A Vast and Magnificent Land — An Illustrated History of Northern Ontario.* Thunder Bay, Ontario: Lakehead University, 1984.

Davies, R.E.G., et al, *The Bush Pilots.* Alexandria, Virginia: Time-Life Books, 1983.

Duncan, David M., *The Canadian People.* 1921.

Edwards, B.A., *Canadian Forces - Sioux Lookout - 30th Anniversary - 1953-1983.* 1983.

Foster, J.A., *The Bush Pilots — A Pictorial History of a Canadian Phenomenon.* Toronto: McClelland & Stewart, Inc., 1990.

Historical Society of Sioux Lookout, *Tracks Beside the Water - Volume I.* 1982.

Historical Society of Sioux Lookout, *Tracks Beside the Water - Volume II.* 1986.

Kiebuzinski, Rae, *Yesterday the River — A History of the Ear Falls District.* Ear Falls, Ontario: Township of Ear Falls History Committee, 1973.

Milberry, Larry, *Aviation in Canada.* Toronto: McGraw-Hill Ryerson Limited, 1979.

Molson, K.M., and Taylor, H.A., *Canadian Aircraft Since 1909.* Stittsville, Ontario: Canada's Wings, Inc., 1982.

Molson, K.M., *Pioneering in Canadian Air Transport.* Winnipeg, Manitoba: James Richardson & Sons, Limited, 1978.

Parrott, D.F., *The Red Lake Gold Rush.* D. F. Parrott, publisher, 1984.

Parrott, D.F., *The Second Gold Rush to Red Lake.* D.F. Parrott, publisher, 1974.

Richthammer, John E. *The End of the Road.*

Russell, Ruth Weber, *North For Gold — The Red Lake Gold Rush of 1926.* Waterloo, Ontario: North Waterloo Academic Press, 1987.

BROCHURES

Bain, D.M., *Canadian Pacific Airlines.* 1987.

Brown, Carson, *The Red Lake Gold Fields.* 1973.

BIBLIOGRAPHY

Campbell Red Lake Mines Ltd., *Campbell Red Lake Mines Limited.* 1986.

Canadian National Railways, *Canadian National Railways.* 1989.

Lake of the Woods Control Board, *Managing the Water Resources of the Winnipeg River Drainage Basin.* 1983.

Manitoba Energy and Mines, *Mining in Manitoba.* 1987.

Ontario Hydro, *Gift of Nature.* 1986.

Ontario Sunset Country Travel Association, *Patricia Vacation Area.* 1986.

Red Lake Publicity Board, *Red Lake District.* 1987.

NEWSPAPERS

Arizona Republic July 1, 1973

The Dryden Observer 1970-1971, 1989

Ear Falls Echo September 18, 1976

Ear Falls Eagle 1984, 1985

Great Lakes Coalition News Fall/Winter 1986

Hudson Newsletter September 1989

Kenora Daily Miner April 11, 1967

Local Express 1989

North Western Explorer June 1988

North Western Explorer September 1989

Red Lake News December 17, 1937

PERIODICALS AND RELATED PUBLICATIONS

Appel, Joyce M., *The Patricia.* 1980.

Appel, Joyce M., *The Work Horses of the North.* 1980.

Cormier, Paul S. *Development of Hydro Electric Generation.* 1987.

Green, Timothy, *Modern Maturity - The World of Gold.* 1988

Nichols, Ron, and Roadhouse, Morley, *Goldpines History Paper.* 1960.

Parrott, D.F., *Red Lake Area Mines Will be Dewatered.* 1987.

Robinson, Ron, *Ear Falls.* Trade and Commerce. 1983, 1985, 1987.

Robinson, Ron, *Red Lake - Gold Capital of Ontario.* Improvement District of Balmertown. 1985.

Sunset Country Travel Association, *Voyageur Travel Guide.* 1987.

REPORTS

Annual British Columbia Minister of Mines Reports re: Granduc Mines Limited. 1953-1957.

British Columbia Minister of Mines and Petroleum Resources Report re: Granduc Avalanche Disaster. 1965.

British Columbia Minister of Energy, Mines and Petroleum Resources, *Minifile 1048-021.* 1989.

Brown, Baldwin Nisker Limited Report. March 1988.

Financial Report, Survey of Predecessor and Defunct Companies. 1985.

Manitoba Energy and Mines, *Sherritt Gordon Mines Limited — Lynn Lake Manager's Annual Reports.* 1945-1954.

Manitoba Energy and Mines, *Sherritt Gordon Mines Limited — Lynn Lake Production Statistics.* 1978.

Ministry of Northern Development, *Ontario Mine Reports Re: Berens River Mines.* 1938-1949.

Newmont Corporation, *Granduc Mines Limited — Engineering and Feasibility Report.* 1963.

Newmont Mining Corporation, *Berens River Mines Limited Annual Reports.* 1938-1949

Ontario Department of Mines, *49th Annual Report.* 1940.

Ontario Department of Mines, Geology of Domes Township Report No. 45. 1966.

Ontario Department of Mines, *Sessional Paper No. 4.* 1940.

Township of Ear Falls, *Ear Falls Progress Report.* September 1983.

Turnbull, R.O., *Winter Tractor Freighting Report - North of Latitude 53.* 1947-1965.

PHOTO CREDITS

Gerald Beckett: p.39.

Derek Bennett: p.35; p.43.

Leo Bernier: p.12; p.17

Evert Cummer: p.156.

U.M. Dart: p.66.

Harry Everett: p.146; p.149.

D.J. Learmonth: p.42; p.125.

C.Elliott Lloyd: p.86; p.133.

Mike Matty: p.57.

National Archives of Canada: p.119.

Mel Parker: p.72.

Provincial Archives of Manitoba: p.98 (C.A.L. Collection 1727); p.151 (Thompson 44).

The Red Lake Museum: p.124.

James Richardson Collection: p.18-19; p.23; p.24; p.30; p.32; p.33; p.38; p.40; p.44; p.46; p.47; p.48; p.50; p.52; p.54; p.55; p.60; p.62; p.64; p.65; p.69; p.71; p.74; p.75; p.80; p.89; p.96; p.97; p.101; p.103; p.104; p.106; p.107; p.112; p.113; p.116; p.122; p.126; p.129; p.136; p.142; p.153; p.159; p.162; p.166.

Allan Sandin: p.59; p.67; p.68; p.79; p.144.

Western Canada Aviation Museum: p.93.

Western Canada Pictorial Index: p.4 (824-24647, UofM Tribune Collection); p.8 (64-1955, The Stark Foundation); p.108 (40-1181, Parrott Collection).

Jack Wish: p22; p.27; p.56; p.81; p.82; p.91; p.100; p.128; p.135; p.157; p.168.

Borys Zayachivsky: p.7.

INDEX

INDEX

I N D E X